KU-617-055

INVEST IN LIVING

HOME POULTRY KEEPING

by

Geoffrey Eley

EP Publishing Limited

Acknowledgements

My thanks for help with the attractive line drawings, essential in a practical handbook of this sort, go to commercial artist Graham Loveland and his sister Judy Loveland, of Wolverhampton.

For technical checking and reading of the proofs, I would like to thank Mr P. R. J. Sharman, Head of the Poultry and Science Department at Kesteven Agricultural College.

Lastly, to Alan Thompson, author of *The Complete Poultryman* (published a quarter of a century ago) and a fellow member of the BBC's 'Backs to the Land Club' which flourished during and for a while after the war when food was in short supply, I owe much of my enthusiasm for poultry as a worthwhile, outdoor hobby and not a little of my know-how.

The *Invest in Living* Series

All About Herbs
Fruit Growing
Gardening under Protection
Getting the Best from Fish
Getting the Best from Meat
Growing Unusual Vegetables
Home-Baked Breads and Scones
Home Decorating
Home Electrical Repairs
Home Energy Saving
Home Furnishing on a Budget
Home Goat Keeping
Home Honey Production
Home-Made Butter, Cheese and Yoghurt
Home-Made Pickles and Chutneys
Home Maintenance and Outdoor Repairs
Home Plumbing
Home Rabbit Keeping
Home Vegetable Production
Home Woodworking
Improving Your Kitchen
Meat Preserving at Home
101 Wild Plants for the Kitchen
Wild Fruits and Nuts

Copyright © EP Publishing Ltd, 1976, 1977, 1978, 1979

ISBN 0 7158 0456 1

Published by EP Publishing Ltd, East Ardsley, Wakefield, West Yorkshire, 1976

Reprinted 1977
Reprinted 1978
Reprinted 1979
Reprinted 1981

This book is copyright under the Berne Convention. All rights are reserved. Apart from any fair dealing for the purpose of private study, research, criticism or review, as permitted under the Copyright Act, 1956, no part of this publication may be reproduced, stored in a retrieval system, or transmitted in any form or by any means, electronic, electrical, chemical, mechanical, optical, photocopying, recording or otherwise, without the prior permission of the copyright owner. Enquiries should be addressed to the Publishers.

Text set in 11/12 pt Univers, printed by photolithography, and bound in Great Britain by G. Beard & Son Ltd, Brighton.

Contents

Health and Pleasure

There is no better value, or more nourishing and concentrated food, than an egg.

Eggs contain carbohydrates, fats and proteins, essential vitamins and minerals —including iron, a vital constituent of healthy blood. Moreover, an egg is one of the cleanest possible foods, protected as it is from all dirt and germs by its shell, and no substitute can be found for it.

For over 2000 years poultry have been kept in Britain and for an even longer span in the Mediterranean lands. Today, domestic poultry keeping is both a worthwhile hobby and an outstanding contributor to the ideal of home self-sufficiency.

Apart from the strong appeal of producing at least some of one's own food, and its contribution to the nation's total supplies, often the responsibility for the care of a domestic poultry unit is given to younger members of the family and this can be a rewarding experience for young people. It fosters a basic knowledge of livestock management, simple biology and the processes of life, and competent record keeping. Even discipline comes into it—keeping poultry is a seven day a week job, with such daily chores as feeding, watering and egg collecting.

Anyone wishing to keep poultry in gardens or on allotment sites ought, however, first to ask the local authority (usually the Environmental Health Department) as to whether or not there are any bylaws relevant to domestic poultry in the area other than the normal regulations in force everywhere covering sanitation, noise, smells and 'nuisance' to neighbours. Making sure of local regulations is particularly necessary in the case of people occupying council owned property.

Even if you have only limited space, as few as six well-managed hens will keep a family in eggs. With the high cost of the necessary concentrated feedstuffs (and, remember, laying birds cannot thrive on household scraps alone) you will not show much cash profit from poultry keeping on a small scale—but you **will** have better quality and fresher eggs than any you can buy, as well as enjoying an absorbing hobby.

There are some 'sideline' profits to be had from domestic hens. Not the least of these is your own supply of poultry manure, a highly concentrated fertiliser for garden or allotment. Neither will the occasional boiling fowl for dinner come amiss when the useful laying life of a hen is over.

In becoming a domestic poultry keeper you will, like the man with a well-managed, high yielding allotment or vegetable garden, be making a contribution to the national larder— around a million 'family units' produce almost a thousand eggs apiece each year.

Basis of Success

Domestic poultry should lay between 150 and 200 eggs per bird each year. This is an average of three or four eggs from each bird in a week—but do not expect their contribution to be as regular as this since the birds usually lay most heavily in the Spring and more spasmodically in the late Autumn. To achieve this level of egg production at home there are four basic principles to be grasped:

Keep only first-class pullets which have been selected and bred to lay (you are unlikely to find these in the market, so buy from a reliable poultry breeder).

Enough properly balanced food to produce eggs.

A poultry house that can be kept dry, is easily cleaned and kept hygienic.

Separation of pullets from any year-lings or second-year birds (older hens often harass young stock and stop them getting enough food).

The way to avoid poultry keeping becoming hard work, or giving you the feeling that you cannot leave them for a weekend, is to plan housing require-ments and other matters carefully from the outset.

Good Housing

The initial cost of a good housing system is unavoidably high, largely because timber is now expensive, but it will be fully repaid in high egg yields over many years—and even if you gave up the hobby the second-hand value of sound poultry houses is always substantial.

Although hens dislike mud and wind they can stand any amount of cold; they will even lay right through the coldest spell if they are properly managed—that is, protected against wind and given plenty of food and unfrozen water. Therefore the essentials of housing requirements are: Dry feet and litter and all the fresh air possible without draughts.

When selecting a site for your poultry house consider the soil drain-age, air movement, location of the dwelling house and the water supply. Remember there is the possibility of such problems as smells, flies, rats and mice. Good soil drainage assures dry floors which will help prevent wet litter, dirty eggs, disease, or other problems.

The poultry house should be located where the prevailing summer winds will not carry smells towards your house, or those of neighbours.

A site on relatively high ground with a south or south-east slope and good natural drainage is best. The location of a poultry house at the foot of a slope where soil or air drainage is poor, or where seepage occurs, is unwise.

If the house is located on a hillside, the site should be graded so as to carry surface water away from the building. In some instances, tile drains will be necessary to adequately carry the water away from the foundation of the building.

In areas of severe winter weather, most of the windows should be in the front of the building; the house should face south to take advantage of the sunlight. The pen will be warmer, litter drier and the birds more comfortable. The closed or back side of the house should provide maximum protection against the north-west wind. Where the prevailing winter storms are from the west, however, the house should face east.

The size of the poultry house will depend upon the type and number of birds to be housed as well as the management system to be used. Various age groups and species of birds require different amounts of floor space for optimum results. In addition to the space needed for the birds, there should always be some additional space for storage of feed, supplies and equipment.

The removal of moisture from the building is one of the main problems in poultry keeping. The moisture in freshly voided manure is as high as 70 to 75 per cent, added to which there is the hens' respired moisture and further moisture in the in-coming air.

Good insulation of the poultry house will provide maximum bird comfort as well as helping to control excessive moisture.

Wood is Best

If you are buying a ready-made poultry house—and there are many good ones on the market, some shown in Figs. 1 to 5—do make sure that the wood from which it is made is not too thin. Wood of this kind warps easily and is unable to withstand extremes of temperature; the result is a poultry house the inside of which is like an oven in summer and a refrigerator in winter.

Whatever type of house you buy or build for yourself, allow between 1 and 2 sq feet of area per bird—even up to 3 sq feet (.3 sq m) if you choose hens of a large breed.

A very common mistake made by domestic poultry keepers is to provide the hens with only one earth or grass

Fig. 1. A practical poultry shed and scratching run

Fig. 2. A useful laying shed for the 'small' poultry keeper

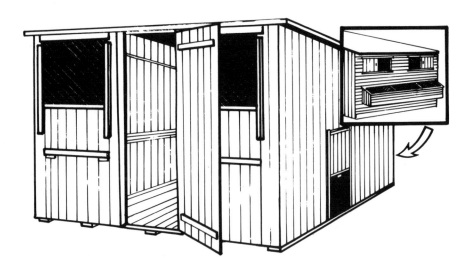

Fig. 3. A typical 3 m × 2 m poultry shed

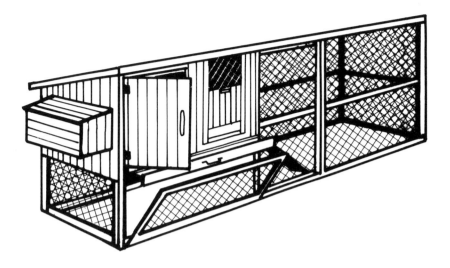

Fig. 4. 'Hen pen' for limited space poultry keeping

Fig. 5. A slatted or wire-floored field laying house

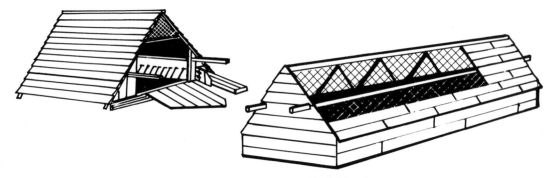

Fig. 6. Examples of movable fold units

run. There should be at least two, to be used and rested alternately since this is the only way to avoid poultry runs from becoming like a quagmire in winter and your birds looking, and indeed feeling, unhappy. If you have enough space—perhaps a small orchard in country districts—the fold units shown in Fig. 6 are very convenient for moving about on grassland.

Where space is limited, poultry can be successfully kept by making use of a garden wall against which can be placed a covered run and small house (see Fig. 7). Such a covered-in scratching area can consist simply of wire netting sides with a roof sloping upwards to the top of the wall.

Advantages of the hen-pen are that it can be easily moved from one place to another and with its covered top, the scratching area is kept dry. The type of pen illustrated can be bought or made very simply, to a size of around 9 to 12 feet (3 to 4 m) long and $1\frac{1}{2}$ yards (1.5 m) wide—room for six or eight birds.

(Here, and elsewhere in this book, convenient rather than precise metric equivalents are used.)

All types of poultry house can be greatly improved by covering the outside with felt—not the inside, as is sometimes done, because this only harbours insects and vermin. Regular creosoting or tarring your poultry house will help keep both it and the birds free from insect pests.

The Roof

The roof is a vitally important matter in poultry housing—it must fit extremely well otherwise you will be in trouble with damp.

If you are making your own poultry house, make sure the roof comes down over the eaves by at least 3 inches (75 mm) to carry off the rain. It is also helpful to have the roof guttered and allow the rain to drain off to a down spout and into water butts (to say nothing of also providing you with a plentiful supply of rain water, so much better than tap water for the garden plants).

Cover the poultry shed roof with felt and put a generous quantity of tar on it once a year.

For permanent houses, corrugated iron lined with matchboarding $\frac{1}{2}$ inch (10 mm) thick, with a layer of felt between, makes perhaps the best possible kind of roof. It is expensive but has the merit of being warm in

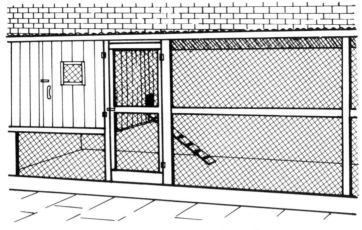

Fig. 7. Lean-to on the wall of a house

winter and cool in summer, and of course it lasts a very long time without attention.

If you decide on this kind of roof, allow a space of about 2 inches (40 mm) between the felt and the iron to allow a current of air to pass between.

Ventilation

Good ventilation is necessary to prevent birds suffering from respiratory diseases. It may sound surprising but poultry in relation to humans and the relative body weight need over twice as much air as we do! Inadequate ventilation causes a strong smell of ammonia and a simple test is a sniff when you feed the birds in the mornings—there should be no suspicion of 'fug'.

It is best to arrange ventilation from the front of the house (see Fig. 8) and if your house is, say, 6 feet (2 m) high, then the lower half or slightly more should be boarded up. Cover the top, open portion with fine mesh wire netting, fastened from the inside. Over this fix a light wooden frame fitted with glass—the idea is that you should be able to lower the glass frame to the bottom of the wire netting in hot weather or raise it nearly to the top when it is very cold. An alternative is to arrange the glass frames to open outwards at a slope, hinged at the top and bottom.

No such special ventilation windows are needed in a lean-to type of poultry house since the permanent open front is sufficient.

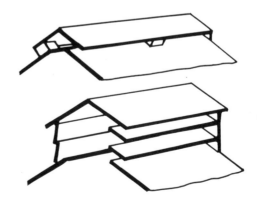

Fig. 8. Outlet ventilators for poultry houses

11

The Floor

There are probably three satisfactory ways of making a floor for your poultry house—boards, cement, or earth and gravel well beaten down.

Well-tarred straight boards 1 inch (25 mm) thick should be used. Fit them close together and inside the foundations of the house itself. If they are closely fitted and nailed across a couple of stout battens, no damp will get in from below. It is **not** a good plan to construct the floor of your poultry house slightly larger than the house itself. Constructions like this, where the house itself stands on the larger floor area, mean that driving rain will most certainly penetrate during the winter.

Cement is the second mentioned type of flooring and while this has some merits—for example it is durable and permanent—the domestic poultry keeper must be careful that he uses sufficient depth of litter to avoid any ill effect that might otherwise accrue from such a hard, cold type of flooring.

Where it is proposed to have earth or gravel as the floor, be sure to mark out the site of the poultry house in dry weather (taking care that the hole lies square) and that surface soil is removed to a depth of about 5 inches (125 mm). The resulting hole should then be filled with good gravel, on top of which sufficient brick ends and similar rubble (clinkers would do) should be well pounded down until the hole forms a hard and level surface.

If a course of bricks is laid all round the outside so that the foundations of your house lie level upon them, and plenty of bedding is used inside, then this type of flooring is entirely satis-factory—except, perhaps, if the situation is very low lying when rising damp could be troublesome.

Nest Boxes

It is a good idea to allow one nest box (Fig. 9) for every three hens. Remember the hen likes to be sociable, although secluded, when she is laying, and the

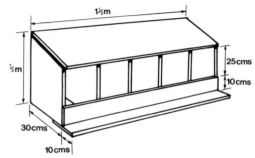

Fig. 9. Nest boxes constructed of plywood or 2 cm timber

more warm bodies already in nest boxes the stronger is the hen's wish to force her way into the box!

Darkness is important. It tends to stop the birds eating the eggs, and the hen naturally chooses herself a dark spot in which to lay her eggs. If the nest box in your poultry house is exposed to full light, then cover it with a hanging piece of sacking.

Always place your nest boxes off the ground—but not at the same height as the perches in the hen house, otherwise the birds will roost in the nest boxes instead of on the perches. Litter the nest boxes with straw (rather than hay), wood shavings or sawdust.

Perches

Many small troubles of the poultry keeper can be traced to improper or defective roosting accommodation. By

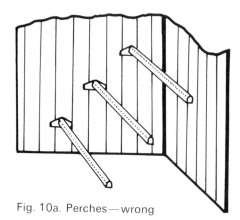

Fig. 10a. Perches—wrong

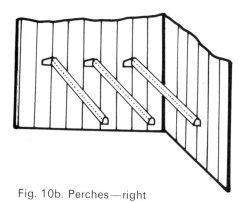

Fig. 10b. Perches—right

giving a little thought to the way in which the perches are arranged and fixed in your house, both loss and worry can be avoided (see Fig. 10). The best material for perches is timber of 3 inches (75 mm) thickness, with the top edges rounded off.

The most common error is overcrowding on the perches. Allow at least 8 inches (200 mm) perching space for each bird and you will not go wrong.

Another fault is to place your perches too high from the floor which leads to cut and sore feet, particularly in the case of heavy breeds, when they alight from such perches.

Never use poles or wood with the bark on. When the wood dries the bark becomes loose and the hollow space provides a grand retreat for thousands of undesirable insects. Neither should you leave any sharp edges if you use purchased timber to make your perches.

Hoppers

These are receptacles for dry mash, corn or grit, usually kept under cover and so constructed that the birds can help themselves and at the same time further supplies keep coming down automatically from the reservoir above. If you use this kind of self-service hopper (Fig. 11) make sure the feeding stuff you are using does not clog up in the upper part of the hopper.
(See section on Feeding, page 28).

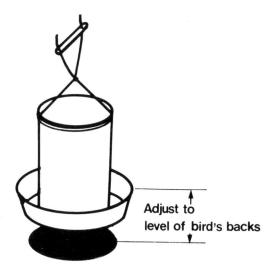

Adjust to
level of bird's backs

Fig. 11. Feed hopper

Building your Own

The house and run shown in Fig. 12 is designed to be built quite easily by the home handyman and to provide ample room for up to 12 well-managed birds. It is a design which has stood the test of time and is eminently practical, and the plan drawings give both Imperial and Metric measurements.

It will be seen from the illustration that the roosting house is raised 2 feet (.5 m) above the ground level. This serves three good purposes: (1) It provides additional space to the run; (2) It does not harbour rats and other vermin as houses do built just off the ground; and (3) The floor of the house provides shade from the sun during the summer.

Another feature is the position of the nest boxes in a detachable unit across the front of the structure. The position is not only easy for egg gathering but it is also the darkest part of the house, which the birds prefer for egg laying.

A door is provided at the side, but this position is optional. If there is not sufficient space for a side entrance, the nest boxes could be transferred to the side, and the door to the front; in which case it would be necessary to alter the general arrangement of the framework by shifting the window to either the right- or left-hand side of the front.

The small trap door between the house and the run is operated from the outside by means of a cord passing through a couple of screw-eyes fixed in convenient places in the side of the

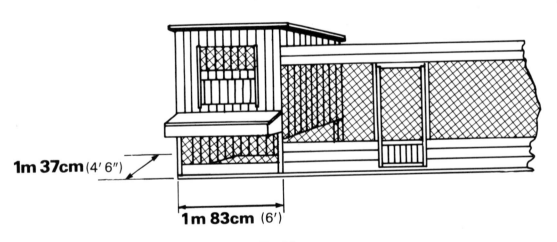

1m 37cm (4' 6")

1m 83cm (6')

Fig. 12.

structure. This simple device saves the bother of entering the run every time the trap door has to be opened or closed. It is kept in the open position by placing the looped end of the cord over a rail.

Size of House

The total length of the house and covered run is 18 feet (6 m); 6 feet (2 m) being the length of the house and 12 feet (4 m) that of the run.

This measurement is an arbitrary figure because the length of the run can either be reduced by about 2 feet (.5 m) or added to, depending on space available.

The covered run is simple enough, although it may be an advantage to add a hood in front, for further protection in bad weather. The height should be about 1 foot (.3 m) less than that of the roosting house, and the slope of the roof made to correspond.

The framework of this part is constructed of 2 inches by 2 inches (50 mm by 50 mm) section timber, and is covered with planed and rebated weatherboards at the back end and part of the front, while the top is covered with $\frac{1}{4}$ inch (5 mm) V-jointed tongued and grooved matching.

The roosting house is made in sections, each of which is framed with 2 inches by 2 inches (50 mm by 50 mm) timber. Tongued and grooved V-jointed matched boards, about $6\frac{1}{2}$ inches (150 mm) wide and $\frac{1}{2}$ inch (10 mm) thick are used for covering purposes.

Tongued and grooved floorboards 1 inch (25 mm) thick, placed across the narrow width of the structure, are used for the floor, and are purposely

left loose for cleaning purposes.

The whole of the structure is supported on two 18 feet (6 m) lengths of 3 inches by 2 inches (75 mm by 50 mm) wood. This method of support will be found much more effective than placing the uprights on bricks, which can sink into the ground and upset the level of the structure. The legs of the house are held in position with small metal angle-brackets screwed to the legs and foundation plates.

Start by sawing up the timber for the framework for the house to the dimensions shown in the drawings (Figs. 13, 14 and 15). Mark each member as it is cut and keep the parts for each section in a separate pile to avoid confusion.

Assembling

Next, assemble the parts. This is quite a straightforward job, as the members are simply butted together and fixed with wire nails.

Having completed each sectional frame, fix the matchboards to the front section. Start at the right-hand end and fix the first board, allowing $\frac{1}{2}$ inch (10 mm) overlapping after having removed the tongue. The overlap covers the ends of the boards on the side section and consequently makes a neat flush finish when erected. A similar overlap is, of course, necessary at the other end.

When cutting the boards for the front section do not overlook the fact that the bottoms are fixed to the third horizontal rail and not the floor rail, as the space between these two members will be occupied by the nest box.

When fitting the boards, it is a good plan to cover the whole area and then

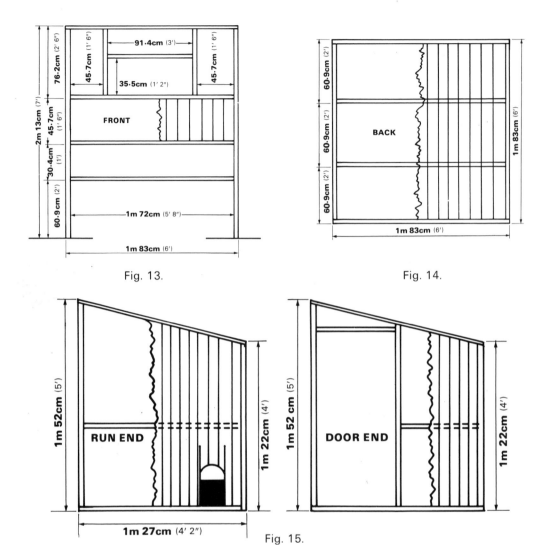

Fig. 13.

Fig. 14.

Fig. 15.

cut out the window aperture with a saw.

Matchboard the back section in the same manner, not forgetting the overlap at each end. The boards of this section extend from the top of the frame to the bottom.

When all the sections are complete, give them a good coat of creosote on both sides and, whilst this is drying, start on the roof.

In the house described, the roof fits on top like a lid (see Fig. 16), the dotted line representing the tops of the walls of the house.

It will be observed that each end board is only partly supported by the longitudinal rafters, but these boards are further supported by the top crossmembers of the sides when assembled.

After the roof boards have been

16

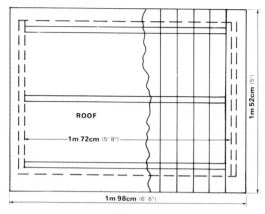

ROOF

1m 72cm (5' 8")

1m 52cm (5')

1m 98cm (6' 6")

Fig. 16.

fixed, remove the sharp edges all round with a plane, to prevent them cutting the roofing material.

Give the roof a good coat of creosote on both sides.

The next procedure is to construct and fit the sliding shutter to the front section; details of this are shown in Fig. 17. The overall size of the shutter is 3 feet 4 inches long and 1 foot 6 inches wide (1 m by .5 m). It consists of matched boards held together by two vertical battens as shown. The shutter slides in rebates or grooves formed by nailing pieces of 2 inches (50 mm) wide batten over pieces half this width, which are then fastened to the front with screws.

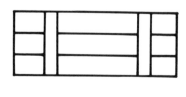

Fig. 17.

The door is of simple construction, consisting of matched boards fastened to three ledges or battens. The length of the ledges should be $\frac{1}{2}$ inch (10 mm) shorter than the total width of the door, to allow it to close against the door-stop.

When Help is Needed

When the holes for the fixing bolts have been marked and bored, the house is ready for assembly, but at least two people are needed to assemble.

When the four sections have been bolted together, test for squareness and vertical accuracy by means of a square and a plumb-bob. If correct, fit the roof, which is secured by driving screws through the bottom of the frame into the plate.

The Roofing Felt

Two 7 feet (2 m) lengths of standard width material will be required, which allows a little all round for turning under the edges and an overlap at the horizontal joint in the centre of the roof. Start fixing at the front edge and work towards the back using galvanised clout nails.

Fold the overlapping portion neatly over and under the edges of the roof and fix it securely by nailing laths on the underside. Finish off the roof by nailing down the three battens—one in the centre and one at each end.

Floorboards

Cut and fit, but do not nail them down, and fix the door to its post and the door-stopping round the inside face of the frame. Make sure that the door closes properly against the stop and that the outside of the door is quite flush with the rest of the surface.

Now make the nest box as shown in Fig. 18 and fasten it to the structure by driving screws through the projecting ends into the two inside faces of the front uprights. The lid should be covered with roofing felt.

The Run

Mark and cut the front and back uprights and mark the exact positions for these on the foundation plates. The back uprights are spaced equally apart, while the front ones can be either arranged to allow the door to adjoin the roosting house or in any other position to suit the poultry keeper.

Fix the uprights to the foundation plates by means of the iron brackets in the same manner as that mentioned for the front legs of the roosting house. The front and back uprights adjoining the roosting house should be fixed to it with screws.

Carefully mark the positions for the tops of the uprights on the long, horizontal top pieces and then fix these members to the uprights with wire nails.

The next thing to do is to fix the cross pieces. These are placed squarely across the horizontals and fastened with nails, the ends being cut off flush with the front after fixing.

Proceed by fitting the rebated weatherboards to the back and side, starting just above the bottom of the foundation plate and working upwards towards the roof. (Test the long edges of the boards with a spirit-level from time to time to check any deviation from horizontal.)

Saw the matched boards for the roof to length, and when securely fastened down give the whole structure a good coat of creosote.

Fixing the Door

The door is the next item to make. The best method is to use morticed and tenoned joints, but halved joints may

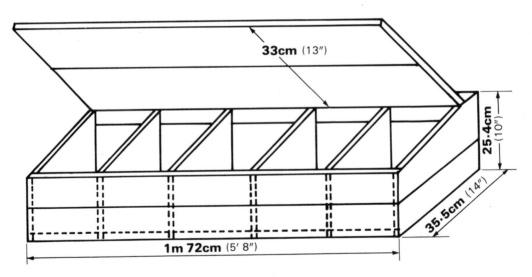

33cm (13")

25.4cm (10")

35.5cm (14")

1m 72cm (5' 8")

Fig. 18.

be used at the top and bottom, and the middle rail fixed either with nails or screws. Another method would be to butt and screw the members together and reinforce the joints by using metal angle-plates.

Having completed the frame of the door, the lower portion should be filled in with matched boards.

Cross garnet hinges should be used for hanging the door which, of course, opens outwards, and a thin strip of wood fitted to the inside face of the door post will prevent the door from being forced inwards and thus straining the hinges.

The wire netting may now be fixed to the inside faces of the framework, etc., and the door, using galvanised staples.

Fix wire netting to the bottom part of the roosting house, fastening a short length of chain to the top of the front sliding shutter and a hook for its reception to keep the shutter closed; screw in one or two screw-eyes for guiding the cord operating the trap door, and fit either turn buttons or padlocks to the doors and nest box cover.

Other sound, well-tried types of poultry housing—all suitable for the home handyman to make himself—are shown in Figs. 19 to 24 inclusive.

Fencing

One of the advantages of a **portable** system of housing is that the necessity for runs is dispensed with, so saving a considerable amount of expense and labour. When, however, space is limited and the fowls must be confined, then the best form of fencing is wire netting fastened to wooden posts. For most breeds of hens, fencing 6 feet (2 m) high is necessary—although, for the heavier breeds which are not so 'flighty', fencing of about 4 feet (1.5 m) is enough.

Do not buy cheap wire netting and also make sure that the mesh of netting

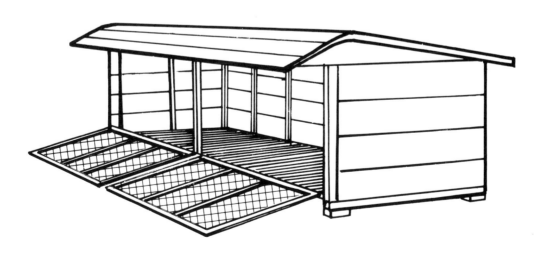

Fig. 19. A useful type of house raised off the ground on bricks or short legs and with a slatted floor

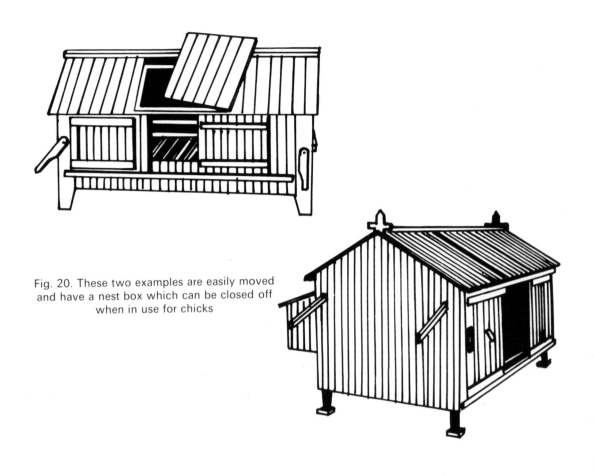

Fig. 20. These two examples are easily moved and have a nest box which can be closed off when in use for chicks

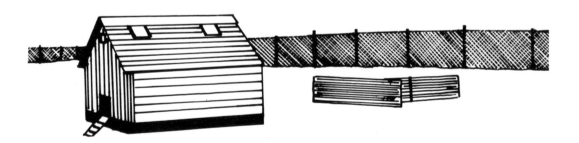

Fig. 21. Type of house used for colony poultry keeping in the open fields

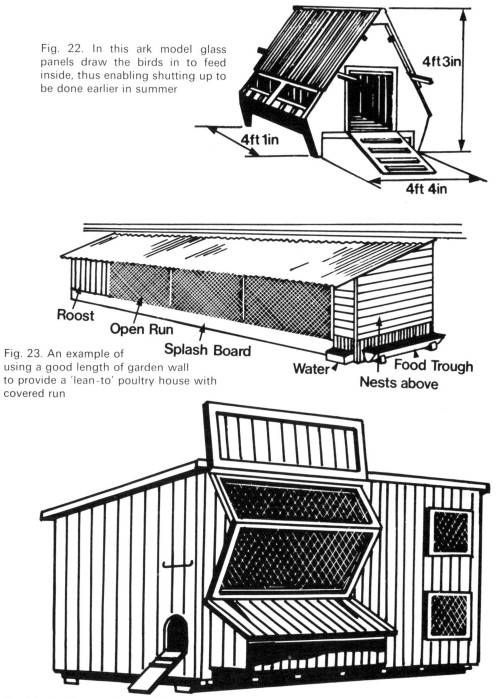

Fig. 22. In this ark model glass panels draw the birds in to feed inside, thus enabling shutting up to be done earlier in summer

4ft 3in

4ft 1in

4ft 4in

Roost

Open Run

Splash Board

Water

Food Trough

Nests above

Fig. 23. An example of using a good length of garden wall to provide a 'lean-to' poultry house with covered run

Fig. 24. Another good type of house; the design of the front allowing maximum light and air so that the house could be used intensively in winter without the birds coming out

is around 2 inches (50 mm). (The term mesh, used in connection with wire netting, means the space in between the wires.)

Keep it Neat

When wire netting 6 feet (2 m) high is used you will need wooden posts long enough to allow for driving them some 16 inches (.4 m) into the ground. The job is made easier if the stakes or posts are pointed at one end (and, of course, tarred or creosoted) so that they can be driven into the ground with a heavy hammer instead of having to dig holes for them.

Spare no pains to make your poultry fencing—and indeed the whole unit—look as neat and trim as possible (there is no reason why backyard poultry keeping need disfigure the surroundings but regrettably this happens, usually because rusty iron sheets, a hotch-potch of posts and other odds and ends have been used).

Dust Baths and Litter

Whatever kind of poultry house you use its tenants must be provided with a dust bath—either inside or out.

The dust bath, as with wild birds, is essential to the well-being of your fowls. Not only does it help to keep them in healthy plumage, but it is also the only means birds have by which they can cool themselves in hot weather or get rid of insect pests.

The term dust bath is used for a heap, box or enclosure of dust or ashes in which fowls delight to wallow, passing the dust between their feathers to their skin—for the purposes described above.

In its simplest form a dust bath is a dusty place in the sun or around your poultry house where the birds can make their own pleasure. The nearest artificial equivalent is a heap of fine, dry soil or—better—a mixture of saw-dust, dry earth, sand or road grit, all kept in place by surrounding boards, and well sprinkled with insect powder.

If you have a fixed poultry house then provide the dust bath in a box, either inside the house or in a hole under the scratching shed. For up to 10 hens the box need be no more than 3 feet by 1½ feet (1 m by .5 m) size.

Litter for Warmth

Litter not only provides a medium for poultry exercise and diversion for the birds in scratching for hidden corn—but it also maintains the birds' warmth and it forms a valuable part of manure for the garden.

Apart from wood shavings, a splendid—but more expensive—form of litter is perhaps peat moss, which is warm and comfortable for the birds as well as being a good deo-doriser. Lay peat moss to a depth of about 5 inches (120 mm) and it will last for six or seven months, provided you turn it over occasionally. And, of course, it forms a most valuable addition to the manure needed for your garden crops.

Use the medium to coarse kind of peat in your poultry house but do remember that as peat is so porous it soaks up moisture like a sponge and in hot weather the air will be filled with dust when the fowls scratch in it. On the other hand, it is no use having peat or any other litter in a damp or leaky house otherwise it will become sodden and therefore bad for the birds.

Collecting Leaves

Using suitable litter for poultry need not be expensive. In the autumn, for instance, you can collect fallen leaves and store them for use dry in the winter—this makes an excellent litter and improves the organic quality of your poultry manure. An alternative is bracken fern which, if cut, harvested and stacked, is quite as good as leaves and, incidentally, it makes a first class litter for your nest boxes since insects and other pests do not seem to like it.

The Way with Straw

Straw is commonly used in small houses and this can be mixed with a little dry earth which seems to help the birds to break up the long straw.

All the foregoing applies, of course, to utility poultry. If you are going to make a hobby of poultry exhibition then you must be more selective in terms of litter—for instance the breeder of feather legged birds cannot allow his birds to scratch about in long litter since this would spoil the foot feather. Neither would the bantam breeder be able to litter his sheds with leaves which would nearly bury his pets!

Intensive Methods

The deep litter system and the built-up litter system are ways of housing poultry when the birds are confined for the whole of their lives and where litter is built up to a depth of between 6 inches and 10 inches (150 mm to 250 mm).

Bricks, concrete blocks, breeze blocks, and wood can all be used for building deep litter houses. If wood is used, it is best to line the entire inside with asbestos or similar insulation material.

The intensive method of poultry keeping is not one often used by the small domestic poultry keeper. One reason for this is, certainly in built-up

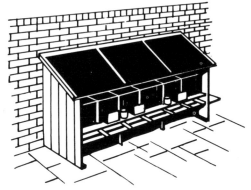

Fig. 26. Besides using laying cages inside another building, such as a shed or garage, a small unit can be placed against a wall (or stout hedge)—provided the cages are well roofed. Ideal for the 'back-yard' poultry keeper without a garden

areas, the problem of obtaining an easy supply of sufficient litter at a reasonable price.

Possibly it would be better, where space is so limited as to prevent having adequate outdoor runs, to consider laying batteries.

Battery Hens

This is the system most frequently used by commercial egg producers.

There are several reasons for its popularity. Floor space requirement per bird is less, thus reducing per bird cost of housing, and the system is labour saving in handling materials.

One method is to provide one cage per bird only, and no heat or other apparatus is needed. A single bird laying cage is usually about 15 inches

Fig. 25. A small-scale laying cage can be a most useful temporary home for laying hens whilst their main house is being 'spring cleaned' or repaired

Fig. 27. A small hen battery

(.36 m) wide, and 18 inches (.5 m) in both depth and height; or you can design cages for more than one bird— with the exception of width, the other dimensions can be the same. The cages (Fig. 27) can be stacked in tiers but not more than four high and in 'units' from eight upwards.

Batteries are, for maximum egg yield, usually housed under cover, but they also give reasonable results sited out of doors, or in covered yards. In either case, hens are never let out. The main construction is of wire netting with a strong, woven netting floor through which the droppings fall to a false bottom in each tier of cages.

Food trough and a water container should be provided in front of each cage but no nest boxes are needed as the eggs, laid on the wire floor, roll out to an egg tray since the floor is constructed with a slope of about 6 inches (150 mm) per yard (1 m). The 'egg cradle' should extend about 7 inches (180 mm) beyond the cage front.

The battery system has its critics. They say it is not natural for birds to be penned up without exercise and with nothing else to do but lay eggs. The evidence shows, however, that birds kept in batteries continue to feed, keep healthy and lay eggs at a normal rate.

Despite this evidence, it must be stressed that even with an amateur's very small battery unit it is essential to allow adequate light and ventilation if the battery is under cover, and to keep the hen's quarters damp and vermin free.

Eggs of very high quality can be produced from birds kept in laying cages, despite the claims of the free-range enthusiasts. But quality does depend very largely on the nutritional value of the food provided—the other factor in egg quality is thought to be inherited. Birds fed on a ration deficient in vitamins A and B, and in such minerals as iodine and manganese, will most likely produce eggs also poor in these nutrients.

One other point in favour of battery systems, whatever its lack of interest for the family compared with poultry running about, is that the lack of exercise means the birds put on more fat and thus become very useful for table purposes.

Buying your Stock

The man with limited accommodation who takes up poultry keeping must be sure to buy the kind of birds most suited to confined spaces.

It is wise to start with three or four months' old pullets—and try to buy these from a breeder of good, healthy stock. Certainly you can find hens for sale in the local market but these are often either past their best laying years or are indifferently bred in the first place. Adequate numbers of eggs depend not only on your feeding methods but on the quality of the stock you buy.

The Rhode Island Red is still one of the favourite breeds among small poultry owners (see also section on suitable breeds, page 56). Sometimes these are pure bred Rhodes and sometimes a first cross—particularly popular hens for eggs being the Rhode crossed Leghorn or the Rhode crossed Light Sussex. The former cross provides perhaps the more active layer, but the Rhode cross Sussex makes a first class bird for the table when her laying time is finished.

Another important point in choosing the right kind of hens is that a cross, between heavy and lighter weight hens, can save on food costs compared with the pure bred heavy, such as the Light Sussex. (See Hybrids, page 60.)

Age of New Pullets

If you want your birds to lay as soon as you get them home then you will have to pay extra for pullets of about six months old. A better way is to buy your pullets in when they are only three or four months old and wait a little while for your first eggs.

It is cheaper still to buy pullets at only eight weeks old—by which time they are well feathered and hardened off—but you will have to feed for a fair time before collecting your first egg.

How to Feed Correctly

On no account delude yourself that a few backyard hens will give you a supply of eggs merely by being fed with scraps of food from your table. At most, domestic waste should not account for more than a fifth of total feed.

On correct feeding depends, in the main, the success of any livestock keeping and poultry are no exception. A hen does not lay eggs for your breakfast of its own free will—in her natural state she would only lay the eggs she wished to sit upon, but man has bred the domestic hen to continue to lay eggs for a much longer period than in the natural state and to do this **she must be fed properly**.

Structure of the Hen

To feed correctly it is necessary to know something of the internal structure of a hen and, in particular, how she **uses** her food.

The hen's internal organs are in three divisions—respiratory, digestive and reproductive.

The respiratory organs are: (1) the nostrils through which air passes and where any solid particles are prevented from passing into the lungs; (2) the trachea, or windpipe, which begins at the glottis or slit at the back of the mouth; and (3) the lungs, situated close up to the ribs between the shoulders.

First of the **digestive** organs is the crop, a bag at the bottom of the gullet (or throat) into which food is received immediately after being swallowed and before passing into the gizzard—a 'stomach' where the fowl pulps up the food for assimilation by the intestines.

In her daily wanderings a hen swallows a mixture of things—corn and seed, grass and other herbage, soft solids like bread and poultry meal, indigestibles like grit.

Why Hens Need Grit

The reason for supplying grit is to help hens digest food—grit in their gizzard acts like teeth for the bird. Grain can be ground by normal muscular action of the gizzard, but not so efficiently, since grit greatly increases the number of grinding surfaces and consequently the food plant cells are broken down thus allowing the fullest action by the digestive juices.

Grit is also a means of supplying calcium, an essential part of poultry diet. Although this mineral is sometimes included in purchased balanced feeding stuffs it can also be given to the birds as soluble oyster shell or ground limestone.

It is a good plan to allow laying birds access to both the soluble and the insoluble grits (granite and flint usually) in separate containers. Insoluble grit can be retained in the gizzard for many months and adult requirements are under half a kilogram a year.

There is no predigestion process

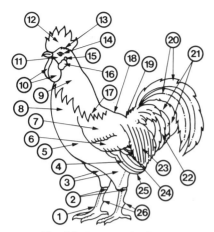

Fig. 28. Points of a fowl

1 Shank or leg	15 Ear
2 Hocks	16 Ear lobe (large
3 Thighs	and fleshy)
4 Primaries	17 Cape
5 Point of heel-bone	18 Back
6 Wing coverts	19 Saddle
7 Wing – bow	20 Sickles
8 Breast	21 Tail (main
9 Neck hackle	feathers)
10 Wattles	22 Tail coverts
11 Face	23 Saddle hackle
12 Comb	24 Secondaries
13 Base of comb	25 Abdomen
14 Cap	26 Spurs

because a hen, unlike us, has no teeth and so all she eats passes into the crop where a softening process is started by a mixture of saliva and water.

The food then passes through the proventricle or stomach—a comparatively small organ in a fowl—and into the gizzard, where the food is ground up by strong muscular walls. From the gizzard the nutritives in food are, as in the human, acted upon by bile (a bitter fluid secreted by the liver and stored in the gall-bladder) and so into the blood.

How Eggs are Produced

The reproductive organs (testes) in the male fowl are high up in the abdomen, near the kidneys. In the female the egg organs are the ovary and the oviduct. There are, in a chick, two ovaries but only the lefthand one develops to maturity—and this is an unevenly-shaped bag, to which are attached the ova, or eggs.

The eggs develop one by one, become detached and slip into the oviduct—a long, twisted tube ending in the vent, or anus, of the bird through which the bird's droppings and eggs pass on leaving the body.

The hen's oviduct is in two parts. In the first portion the white (or albumen) is deposited around the yolk of the egg and in the second part of the oviduct the shell is made. To pass through the first portion of oviduct an egg takes between 3 and 5 hours, through the second 15 to 20 hours. The egg is then complete and ready to lay.

(How much more we appreciate an egg when we know something of the long and wonderful process which has gone into its making! Incidentally, the different colours of eggs are due to pigment deposited from the blood.)

A Balanced Diet

Your birds need feedstuffs containing not only protein, fats and carbohydrates, but also vitamins and certain minerals. This is called a 'balanced' diet.

There is no particular merit in buying the individual components of a balanced diet because it is not only easy enough to purchase the **complete** food from a local corn merchant or town pet shop—but this way you will be certain that the 'mix' is correct. (You will find corn and agricultural

Fig. 29. The development of an egg

(a) The ovary, showing yolks or 'ova' in different stages of development, held together like a cluster of grapes. As each yolk or ovum grows to its full size it drops off and passes through (b) the mouth of the oviduct; (c) the open mouth receiving it. Immediately upon entering the oviduct it meets the spermatozoa and becomes fertilised. (d), (e) and (g) show the egg in different stages during its passage to the outer world. It passes with a rotatory motion, layers of albumen being twisted around it, and then first the shell-membrane is formed and afterwards the shell itself. It takes from 3 to 4 hours for the egg to pass down to point (d), then 16 to 18 hours for the rest of the journey—so that often there are two or three eggs together in the oviduct. The point (f) indicates the last convolution of the passage, and there the wall of the oviduct is thickest and strongest—the shell having practically become hard; (g) shows it about to pass into the cloaca, and (h) the cloaca communicating with the outer world through the vent

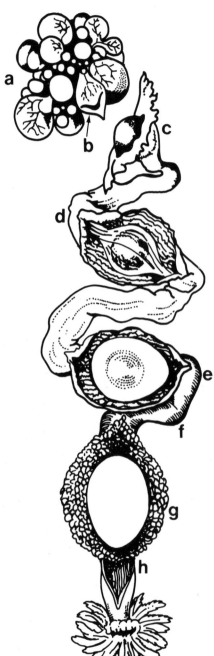

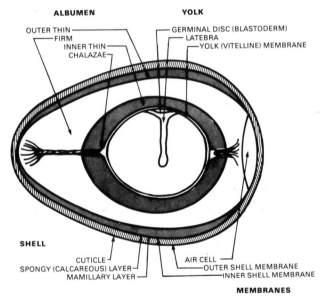

ALBUMEN — OUTER THIN — FIRM — INNER THIN — CHALAZAE

YOLK — GERMINAL DISC (BLASTODERM) — LATEBRA — YOLK (VITELLINE) MEMBRANE

SHELL — CUTICLE — SPONGY (CALCAREOUS) LAYER — MAMILLARY LAYER

AIR CELL — OUTER SHELL MEMBRANE — INNER SHELL MEMBRANE

MEMBRANES

Fig. 30. The structure of an egg

merchants listed in the yellow pages of the telephone directory.)

Poultry need such quantity of food as will maintain the fowl's body, plus enough to provide a constant supply of eggs. The choice of feedstuff is between mash, pellets and grain.

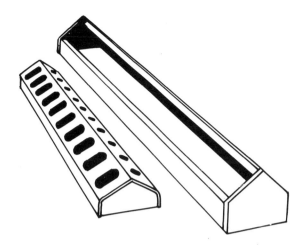

Fig. 31. Feeders

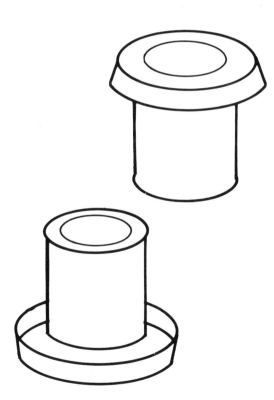

Fig. 32. Homemade watering can, made out of gallon oil can and pan

Bigger, 'heavy' type of poultry eat more than smaller sized birds and all poultry eat more at one time, or one particular day, than at other times and for this reason it is difficult to give a precise measure of food required. However, for the average domestic unit 1 lb. (500 g) of pellets fed morning and late afternoon is quite sufficient for a flock of eight to ten birds.

If you use the 'hopper' system of feeding (see page 13) then, of course, the birds will help themselves. Good models of poultry feeders, and a home-made water receptacle, are shown in Figs. 31 and 32.

One System Only

'Mash' is the term applied to any balanced mixture of meal, while pellets refers to foods that have been ground and compressed into cylindrical form. Pellets are certainly very easy to feed and the size of these varies according to the age of the poultry. 'Crumbs' is food in a granular form.

Whichever system you adopt, use it regularly if you want the best results. Changes in type and method of feeding only result in the birds being disturbed and that usually means fewer eggs.

Certain wild fruits and nuts can be fed to your hens (Fig. 33).

Pellets Easiest

Partly, the type of food you choose depends on the system of poultry keeping. For instance, if you are using the intensive system of housing then dry mash is probably the best because the process of eating keeps the birds more or less busy when they have nothing else to do! On the other hand, pellets are the easiest system of all since they are so easy to handle, are

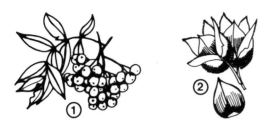

Fig. 33. Nuts and berries that can be fed to fowls

1 Rowan or Mountain Ash	3 Sweet Chestnut
2 Beech	4 Horse Chestnut

Wet Mash

The merit of wet mash feeding is that it allows the maximum use of vegetables and kitchen waste which can all be mixed up with the mash—but you must have ample trough space to allow all the birds to feed at once.

It is a mistake to think that adding water to a well balanced dry mash will necessarily either add to its nutritive value or its appeal to your poultry.

Wet mash feeding, whatever its merits, is also more laborious a task for your wife or whoever else has the job of feeding the poultry whilst you are at work.

Different poultry foods absorb different amounts of water and you can only judge this at the time of actually mixing your mash. If you decide on this method of feeding, the rule is to make your mash moist but crumbly—not sloppy. If the mash breaks up easily when you pass a hand through it then that should be just about right.

Cooked Foods

Potatoes, household scraps and all swill must be thoroughly cooked before you feed it to poultry.

The main reason for this advice is to cut out the possibility of disease—but another advantage is that the birds will find your scraps more digestible when they are cooked.

The nutritional value of an egg is to some extent influenced by the diet, although the general protein and fat content remains fairly constant. The main point to watch is that whatever system of feeding you use the birds must have the right vitamin and mineral content. This is why in the long run it

entirely clean and their cost is only slightly above that of dry meal.

Pellets are particularly popular with the domestic poultry keeper having a small outdoor poultry run because not only are the birds able to 'fill up' quickly but this very fact also leaves them maximum time for foraging in the run, the orchard, or whatever available land they have.

Pellets are also suitable for use in laying cages.

One obvious advantage of pellets is that the almost inevitable waste from dry mash feeding is avoided. A hopper full of pellets is as good a way of leaving your birds to fend for themselves for a few days as you will ever find, and neither will pellets clog the hopper.

is safer to buy ready-mixed foods from the shop or corn merchant, even if it is slightly more expensive than mixing your own feeding stuffs.

Loss of Appetite

Do watch the birds daily habits.

Just like us, they sometimes lose their appetite—and there is usually a reason for this. An instance of this arises sometimes when the system of dry mash feeding is used. Unless the mash is stirred the birds may well find it stale and will not show much interest in adequate feeding. Never just put new food on top of old—not only will the birds not find it very appetising but the mash underneath can become mouldy and cause nutritional problems.

Unless you are using the 'self-service' hopper system (which means food is constantly available to the birds) it is best to feed adult stock only twice a day. Feed times should be as early as possible in the morning, as soon as the poultry house is opened up, and each late afternoon one hour before dusk—or in summer time about 6 pm. Hard corn is ideal for the evening meal whatever food programme you follow, but more so if the birds are on a mash diet.

Good Management

A common mistake among beginners is to assume that the more fowls they keep the better their chances of plenty of fresh eggs. Even those who have been successful with, say, half a dozen hens, want to double the number in the belief that they can double the yield—and all this without increasing their poultry house accommodation.

Another common myth is that the birds will be better if a cockerel runs with them. Not only can this be a noise nuisance in built-up areas but, in fact, a male bird has no influence whatever upon egg production.

The only good reason for keeping a cockerel is when you want to produce fertile eggs to hatch out your own replacement chicks—otherwise a cockerel in the hen yard creates a lot of noise and eats a lot of food!

Yet another mistaken notion is—more food equals more eggs. The result of this misguided kindness is that the birds soon become too fat to lay and develop liver complaints.

These mistakes are especially regrettable because a few fowls will pay you handsomely in eggs if your management is sensible. What, then, are the main ingredients of success for the amateur poultry keeper?

Success depends quite largely on whether you really **like** poultry keeping!

In other words, it is a matter of personal attention, and cleanliness is of the greatest importance where fowls are kept in small houses. Cleanliness, and being scrupulously careful not to leave traces of feeding stuffs around sheds, will also help to prevent rats and mice becoming a pest.

Stress has already been laid on the necessities of good housing, including the space to be allowed for each of your laying hens, and exercise is a good way of keeping hens occupied. This is why some type of covered-in run is essential, otherwise the birds will become lazy in bad weather, staying in the poultry house to avoid the wet and to avoid the quagmire which an open run soon becomes in bad weather.

Water Supply

The importance of a good water supply cannot be exaggerated.

Poultry are naturally thirsty and will drink poor water if they get the chance, so always make sure that your birds have not only enough water but that it is **clean**.

Quality of Eggs

Spicy and other foods that tend to hasten production unduly may also lead to problems including eggs which are double yolked, soft shelled or mis-shapen.

Incorrect feeding can be the cause of very small eggs, one perfect egg being included in a second shell; or the production of the albumen (that is the white of the egg) and the shell but without any yolk.

Eggs with an unusually pale yolk are often the result of an anaemic condition in the hen brought about by under-feeding, confinement which is too close, or an insufficient supply of green food and grain.

Starting Them Laying

With the price of eggs always high in the last three months of the year it pays the domestic poultryman to aim at eggs during this period of the winter when eggs are so much less plentiful than in the spring and early summer.

To do so, the following principles must be followed:
1. Buy in, or rear, pullets hatched in March—April.
2. They should be brought into lay by Michaelmas (29 September) and, if backward, fed extra protein food in the form of fish or meat meal.
3. Housing must be draught-free and dry.

The great mistake is to have only half-grown pullets when the cold weather starts.

Winter Lighting

By artificially lengthening the hours of light during winter days, egg production can be considerably increased. For maximum winter production a pullet is said to need 14 hours of light a day—but one should remember that artificial lighting will not increase a bird's total **yearly** output of eggs; she simply lays more than she would normally do in the dark winter days at the expense of output from spring onwards.

A combination of morning and evening electric lights is the usual method, and by using a time switch the poultry keeper can avoid the chore of getting up early enough in the morning to switch on whilst it is still dark (and turning off each night). When the natural day lengthens to 14 hours again, the lights can be discontinued.

Lamps of 40 to 60 watt are powerful enough and one light fixture is adequate for the average sized domestic poultry house.

When electric lighting is first installed it may be necessary, for the first week or two, to catch and place the birds on the perches if they do not get on the roost of their own accord when it is time for 'lights out'. However, the birds soon get used to the idea—but do not forget they will expect (and need) extra rations if you put them on 'double shift' work!

Replacing Birds

There can be no doubt that the most economical plan is to clear out your old poultry every other year and then start again with fresh point-of-lay pullets (that is when they are about six months old), or even younger birds. The younger the new pullets, the cheaper—but you, rather than the breeder, will face the cost of feeding them until they are old enough to start laying.

The old birds to be dispensed with —or 'culled' from the flock as it is known—will make excellent boiling fowl or they can sometimes be sold live in your local market.

It certainly does not pay to keep hens after their second laying season; their feed costs may well increase and the birds lay fewer eggs. There is a temptation to the amateur—particularly the family man—to keep hens too long because some of them tend to become children's pets!

Every 12 months, usually starting in early autumn, hens will moult. This is the term applied to the casting of their feathers and a rather sorry sight they look—so the time to sell your older birds is **before** they moult; if, however, they are destined for 'the pot' then looks will not bother you!

Coping with the Moult

A common cause of indifferent winter laying is that the hens have not been looked after properly during the moult. Improper care at this time makes the hens backward and unable to withstand winter conditions. Where, on the contrary, hens are well looked after, properly fed and housed in warm and dry conditions, they quickly recover from the August-September moult.

A healthy yearling hen, one not too fat, takes no longer than about six weeks to change her feathers, whereas an old hen takes two and sometimes even three, months to complete the change. If the young hen, therefore, starts about the first week in August, she has finished by the middle of September, which gives her plenty of time to recover her form before the cold weather sets in. Such a hen should be ready to start laying again during the first half of November.

Some varieties of hens take much longer to pass through the moulting period than others—the laying or non-sitting breeds, such as the Leghorns, being generally at least a fortnight quicker than heavy breeds such as the Light Sussex or Rhode Island Red.

Restricted Feeding

During the first part of the moulting period food supply to the birds should be poor in quality and quantity to encourage the feathers to fall as quickly as possible. In fact, commercially, poultry keepers deprive the birds of food and water for up to 24 hours. However, as soon as the new feathers are being formed, make sure you **change back to concentrated layers' food** (see the chapter on How To Feed Correctly).

As moulting takes place during the summer, there is nothing very special to note regarding the housing of your birds. Just make sure they are warm at night and free from draughts.

Partial Moult

The beginner can often be troubled with an outbreak of what is called neck or partial moulting.

It may not last very long, nor spread into a complete moult, but it is enough to cut down your supply of eggs for a month.

Partial moult usually happens during the early part of the year, say January or February, and may be a form of nature's protest against over forcing. For instance, the user of winter artificial lighting (see page 35) who has made his birds put in too much 'overtime', will probably get such a moult among his stock—the same thing happens to the novice who, having obtained good results, tries to force things still further by adding fish or other high protein to his ration.

There is no doubt that the least likely bird to indulge in this partial change of plumage is the pullet who has been hatched in March or April and that has

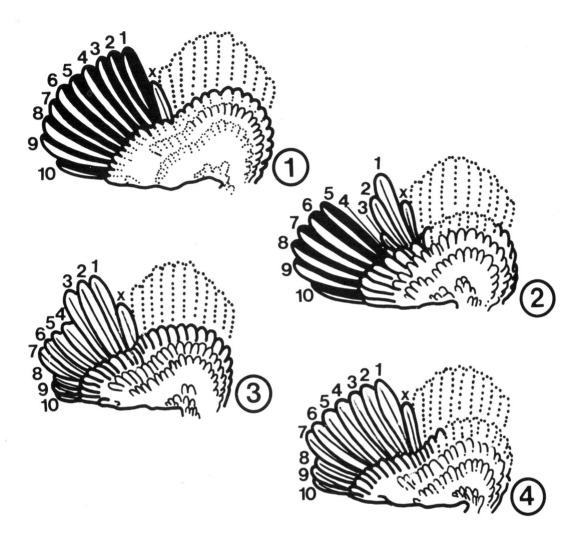

Fig. 34. Wings during different stages of moult. (1) shows the 10 old primary feathers (black), and the secondary feathers (broken outline), separated by the axial feather (x). (2) shows a slow moulter at six weeks of moult, with one fully grown primary and feathers 2, 3 and 4 developing at two-week intervals. In contrast (3), a fast moulter, has all new feathers. Feathers 1 to 3 were dropped first (now fully developed); feathers 4 to 7 were dropped next (now four weeks old); and feathers 8 to 10 were dropped last (now two weeks old). Two weeks later (4), feathers 1 to 7 are fully grown

not been forced on unnaturally in any way, but just allowed to grow and develop in the open.

Some of the Problems

Feather Eating

Few of the troubles that beset the poultry keeper are more annoying than feather eating. It is not a disease, merely a bad habit—but it is most difficult to get rid of.

Often one old hen will start it and she teaches the trick to all her colleagues in the flock. The necks and rumps of every bird soon become denuded of feathers and they present a very sorry sight.

The trouble usually starts in confined runs, presumably because the birds are idle. However, it is also partly a craving for something not supplied in the birds' food and more often than not this is found to be green food.

One of the best remedies, obtainable from chemists or corn merchants, is a very strong solution of quassia, a bitter substance sometimes used as a tonic in medicine, which can be mopped onto the birds' feathers without harm. A cure can also be effected by using proprietary brands of anti-pecking powder. A good plan is to hang up cabbages in the poultry run just above the birds' heads so that they have to jump up in order to peck them —exercise of this sort, clearing the blood, provision of scratching exercise, and similar means of keeping the birds busy, will do more than anything else to stop the nuisance.

Egg Eating

Another bad habit is egg eating—often acquired in the first instance as a result of birds producing eggs without shells or dropping an egg from the perch when roosting. In either case, the hen or hens naturally enough investigate the character of the contents and, having once tasted an egg, they will do it again!

If you always allow your fowls free access to a good supply of broken oyster shell and fresh green food, this will largely prevent egg eating ever starting. This is another instance where poultry kept with an adequate run are better than those in too confined a space—the theory being that a busy hen seldom makes a nuisance of herself.

Egg Binding

This arises principally in two ways— either the oviduct is too small to allow passage of the egg or the egg has become broken and will not therefore slip out properly.

The oviduct of a pullet when she first begins to lay is but a narrow passage and the first few eggs will not find their way through without causing the bird a little pain and difficulty. Most cases of egg binding can be relieved by holding the hen for a little time with her vent exposed to the steam from boiling water—this softens the fat and the pelvic bones, and if the bird is then given some olive oil and placed in a comfortable strawed box she will usually expel the egg within a couple of hours.

If you have not kept poultry before, you may be rather worried when you find your first eggs to be of an odd long and rather narrow shape and possibly streaked with blood. There is, in fact, no need to worry and the reasons are obvious: The resistance of the narrow passage causes the long shaped egg and the rupture of small vessels as it forces its way through accounts for the blood streaks.

Prolapse

This condition, sometimes termed 'down-behind', is fairly common and it is one in which a proportion of the hen's organs protrude. It is caused by straining and it most often happens to hens that are very fat or have been overlaying.

The only thing that can be done is to clean and then smear the part with a lubricating jelly and carefully push it back.

Having done this, put the hen away by herself for a week or more, only giving a moderate amount of food in order to check production of eggs for a time. If protrusion of the intestines recurs then wash or sponge it each time it appears and gently push it back again. Perseverance usually effects a cure.

The Common Cold

This is caused by a virus, quickly spread by overcrowding the birds or in a house having poor ventilation—and sometimes the birds are prone to colds simply through bad feeding.

Egg production falls badly, the poultry seem to lose their appetite, and they sneeze frequently with mucous running from their eyes. Hens do not generally die from this complaint and it usually passes off in a few weeks.

If the colds are only slight, a little sulphate of iron in the drinking water helps—30 gms (about 1 oz) dissolved in $\frac{1}{2}$ litre of water (about 1 pint) and from this solution use two tablespoonsful in each gallon of drinking water.

For bad cases of colds in fowls, the affected birds should be isolated and given the drug Aureomycin, added to drinking water. The drug is obtainable through veterinary sources only.

Roup

If neglected, the common cold soon develops into roup. In 'wet' roup there is an offensive discharge from the breathing tubes but the complaint—although very contagious—can be cured by using pure sulphate of copper. Dissolve 30 gms (roughly 1 oz) in about $\frac{1}{4}$ litre of water ($\frac{1}{2}$ pint) and use the fowls' drinking water at a rate of two teaspoonsful to each $\frac{1}{2}$ litre (1 pint) of water. If many birds are affected, isolation is useless and the above remedy should be doubled in strength and given to the entire flock.

'Dry' roup is a similar condition, but with comparatively little discharge.

Cramp

This is a common ailment of growing stock, arising from imperfect circulation or through damp and cold conditions underfoot. It affects the legs, causing the bird to squat down helplessly, and the hen's toes are often contracted.

The remedy is a change to better conditions—long straw in a dry pen is excellent—and regular rubbing of the hen's legs with a penetrating liniment or even turpentine.

Liver Disease

The cause of this disease, with a heavy mortality rate, is a tuberculosis bacillus which attacks the liver and neck glands in poultry (it is not the same bacillus which causes tuberculosis in man and in cattle).

Lameness in one leg is a preliminary sign and later on a thinning of the flesh on one or both sides of the breast-bone (hence the disease being called 'going light').

A mixture of Glauber and Epsom salts is sometimes given, or used as a preventive, but killing and **burning** of the carcasses is the only sure way of ridding the poultry unit of this disease.

Fowl Cholera

This is a bowel complaint, similar to enteritis, which is both quick in action and fatal—a hen appearing to be normal one day can be found dead the next, usually with greenish-white excreta adhering to the feathers around its vent.

It is caused by a cholera bacillus and is very contagious. The stricken fowl is extremely thirsty before death.

Affected birds must be killed and burnt. The excreta should also be collected up and burnt, and the ground covered with lime. Any birds not affected should be removed to fresh pens.

Preventive drugs are available to add to poultry drinking water.

Newcastle Disease

Another killer, usually referred to as Fowl Pest. One reads a lot about it in the newspaper, when outbreaks occur on farms, but it is fortunately not often met with among domestic flocks—even so, vaccination as a preventive is a good idea, and it is as well to know what are the symptoms of Fowl Pest.

There are acute and less acute forms of the disease; and a more chronic type which shows few symptoms other than those associated with the common cold.

In the acute form, green or greenish-yellow diarrhoea is usually the main symptom. The birds are depressed and tend to gather in a corner of the run displaying marked lack of energy. The less acute type usually shows itself with a watery discharge from nose and eyes.

Crop Bound

One of the commonest of all poultry complaints is congestion of the crop.

For some reason or another the food in the hen's crop cannot be persuaded to pass on and consequently becomes unwholesome, fermentation being set up. The hen will be seen to have a full crop even before the morning feed.

The main causes are either an obstruction (sometimes a twisted ball of long grass) and indigestion, or general debility. The best cure is to give the hen a drink of warm water to distend the crop, which should then be softened by rubbing gently with one's hand. Hold the bird with its head downwards, squeeze her crop and often the liquid will run out. Finish the treatment by letting the hen drink freely, preferably warm water containing a little Epsom salts and bicarbonate of soda (one teaspoonful is enough). Put the hen in a dry pen, and do not feed her for 24 hours.

Pests and other Enemies

In previous sections on housing and poultry management, stress has been laid on the need for cleanliness. One of the main reasons for this is to avoid trouble from pests and various insects, the most common of these being a parasite called Red Mite (see Fig. 35) and the Hen Flea (Fig. 36).

Both these tiny parasites feed on the birds at night and are so frequently over-looked until their damage is severe. During the day the mites hide in crevices, perch sockets and nest boxes —if you suspect or see signs of infection you must thoroughly clean and disinfect the entire poultry house. (Incidentally, the easiest way of finding out whether there are Red Mite is to examine your poultry house by torch-light at night—the presence of Red Mite is also shown up by 'salt and pepper' markings on the perches and elsewhere.)

When you are cleaning up the poultry house for this trouble it is quite effective if you spray with either paraffin or creosote. Another useful point to remember is to avoid bringing in parasite infection when you buy secondhand equipment—always clean it thoroughly before use.

Apart from the fox, which is the main worry to the poultry farmer in the countryside, the amateur is most likely to be troubled, particularly if he lives in a town, either by cats or rats.

As far as cats are concerned, it is best to make sure your stock is properly

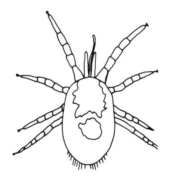

Fig. 35. Adult Female Red Mite (magnified 45 times)

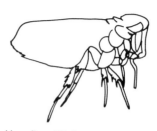

Fig. 36. Hen flea (Pulex avium). Actual size about 3 mm

protected with the use of adequate wire netting. There is, in fact, not much else you can do.

Rat Proof

Rats are, however, another matter. Again use small mesh wire netting plus consistent poisoning whenever rats are seen to be about.

Gardens previously free from rats will often attract the pest when poultry are kept—no doubt brought in by the smell of good food. It is important to keep your feeding stuffs in secure rat-proof tins (a new dustbin is ex-cellent for the job).

Rearing Chicks

It is a fairly safe bet that if the domestic poultry keeper is a family man then sooner or later the children will want to see some baby chicks.

If this is the case, and none of your well-managed hens has gone broody (the term given to a hen who wishes to sit and, with or without eggs under her, stays put in a nest box), a friend might well lend you a broody bird under which you could place 12 to 15 fertile eggs. It is an intriguing side to poultry keeping—but it demands understanding and considerable care. The housing needs of a broody hen and her chicks are explained in the set of drawings shown in Figs. 37 to 44.

The Broody

The first thing to remember if you are going to sit a broody hen is that you must make some concession to the bird's natural instincts—to take a broody hen from the nest of her choice and place her down upon eggs shut up in a box you have made yourself is, from the bird's point of view, not very appealing.

Try to introduce the broody bird gently to your selected nest site, preferably in the evening, so that by the following morning she may be a little more reconciled to her artificial quarters.

It is a good idea to place a few china eggs in the nest to begin with and, after the sitting hen has once been out to feed and has returned to settle down

Fig. 37. When a layer is still on the nest at dusk it can be assumed that she is broody

If the broody hen is not wanted for chick rearing, she is placed in a coop with slatted floor—the discomfort puts the bird off further sitting

again on the china eggs, substitute the real eggs for hatching under the bird at night.

Gentle Handling

Good management of sitting hens requires gentle handling throughout the process of incubation. Each day give the hen enough time for feeding, exercise and scratching about in a dust bath. The hen should be free, and off

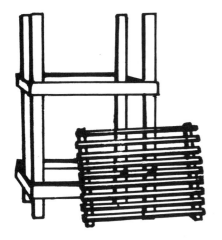

Fig. 38. Handy back garden coop. As can be seen (together with Fig. 39) the maximum use can be made from a coop by making it go with a stand and slatted floor

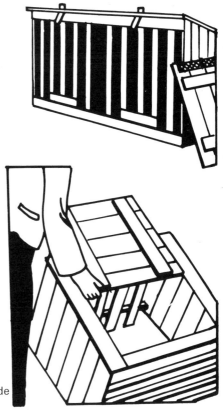

Fig. 39.

Fig. 40. The conventional coop is fitted with a floor to slide out

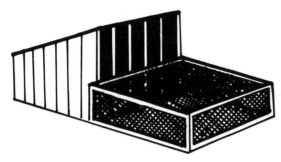

Fig. 41. For mother hen with her chicks

Fig. 43. A fold-type unit used for hens with chicks on a small poultry farm. The outfit above shows a further variation in style of coop

Fig. 42. The receptacles on the top of the run are for grit and water. For a start it is necessary to feed the chicks in the coop above where the hen will teach them to eat

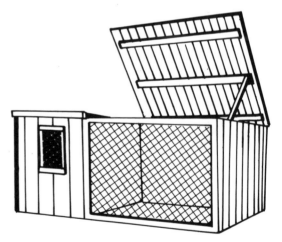

Fig. 44. A double coop is very useful for a hen and chicks. The roof opens up to permit feeding and watering

the nest, for about a quarter of an hour each day.

If the weather is very bad, give your hen some scratching material in a shed and scatter the corn in there to induce her to take enough exercise— and here comes a warning; on no account let your hen re-enter the nest box until she has evacuated, otherwise she will almost certainly foul her nest and the sitting eggs. Should she by chance do so, or soil her feathers by means of a broken egg, then you must clean up the bird.

Start of Incubation

When incubation begins the hen often endeavours to sit tight on the nest, ignoring the necessity for food and exercise. When this happens you should lift the hen gently from her eggs and place her down carefully outside the nest box and within sight of the food and water.

When a hen, whose previous behaviour has been normal, shows a strong disinclination to leave the nest within a day of the due date for hatching, then do not disturb her. The hen's action at this time is directed by her natural instinct regarding the immediate requirements of the nearly due chickens within the shells.

Completion of the development of the embryo varies a little but normally the perfected chick should be ready to break the enclosing shell sometime during the 21st day of incubation. As soon as shell chipping has begun then feed and water the hen without taking her off the nest; any chilling of the eggs at this time is harmful, sometimes even fatal.

Birth of the Chicks

At actual hatching time it is best to 'leave well alone'—like any other mother, the hen knows well enough what to do.

Usually, from the start of the chipping it will take between six and 10 hours for the chick to emerge completely from the shell. All that you need do is to remove the empty egg shells by placing your hand beneath the hen during the hatching process.

If there should be undue delay in the chick emerging, all that can be safely done consists of the further breaking of an already chipped shell and the gentle tearing of the shell membrane if this is unusually dry or tough. But as a general rule, do not interfere.

Transferring to Coop

Before they get their first meal after incubation, chicks and mother should be transferred to a coop. Let the hen examine the coop and settle down in a brooding attitude before you let her have all the chicks—and at this stage dust her with insect powder to free her from insect pests.

When a hen, after sitting tight for two or three weeks, has finished hatching her brood she deserves a good feed of maize and wheat before taking up again her maternal vigil.

Artificial Rearing

The Incubator

An alternative to using your own broody hen is to buy fertile eggs and use a small incubator. There are several of these sold by the poultry appliance makers and, provided you follow the maker's instructions and the rules below, you should be successful:

Use only well-shaped, good sized eggs less than a week old, marking the date on them when placed in the incubator. Do not load the incubator with too many eggs.

Situate the incubator free from draughts and from sunshine and heat it running at about 40°C.

Reduce this temperature two or three degrees each week until the chicks can do without artificial heat at six to eight weeks old.

Throughout incubation keep the water tray in the incubator topped up so that the eggs have adequate moisture—and turn the eggs, in opposite directions, three or four times a day.

At seven and 14 days check for fertility, removing any infertile eggs which will be 'clear' if held to the light. As soon as the chicks begin to chip the egg shells at or about the 19th day, leave the incubator alone. On the 21st morning, inspect the interior and see that all is well with the hatched chicks —but do not remove or assist chicks from the shell. For 24 hours leave the chicks undisturbed, after which they can be removed in a warm, protected basket to the brooder for rearing.

When the chicks are hatched they can either be transferred to a brooder (see below) or put under a hen which has been sitting for three weeks or so and has some chicks of her own. If you try this method of rearing you must gently place the new chicks under the hen whilst she is on the nest, but the incubator-hatched chicks must be of the same colour as the hen's own chicks, otherwise the hen will probably kill the 'foreign' chickens—poultry have a keen sense of colour, even if they cannot count!

The Brooder

This is an appliance for the artificial rearing of chicks hatched in an incubator.

A brooder is made with a metal or wooden floor and sides, with a way out for the chicks at one place only. Although there are plenty of good brooders on the market the handyman can easily make his own provided he remembers to have three parts in the brooder—one for sleeping, another for feeding and exercise, and a third as an outer run on grass, but this must be covered and have a wire netting front to let the light in.

In winter time your chicks will remain in a brooder until they are about six weeks old. The brooder can be heated either by an infra-red lamp (see Figs. 45 and 46) which is cheap and effective to operate, by an electric heating element

Fig. 45. An infra-red brooder

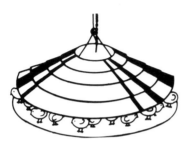

Fig. 46. A hover-type brooder

or by an oil lamp. Keep the heat in the brooder around 35°C, reducing this two or three degrees each week. At six to eight weeks old your chicks can do without artificial heat.

Feeding New Chicks

For the first ten or 14 days chicks are fed on dry chick crumbs or pellets, which is the normal powdery mash processed into small pellets, or fragments. This is available from your pet shop or corn merchant—or, if you have a grinder, you could make your own from the following seeds: Canary seed (3 parts), millet seed (2 parts), groats (also 2 parts) and one part each of hempseed and finely cracked peas. After about 10 days, add an equal portion of finely cracked wheat. A week later still, add a little small maize and broken chicken rice.

For either artificially reared chicks, or those under a hen, 'little and often' is sound feeding advice—at least every three or four hours—and a good plan is to give the morning meals soaked in cold water, but feed dry for the remainder of the day. At all times leave the birds an adequate fresh water supply.

When your chicks are about a month old they should be given **alternate** meals of soft and hard food (such as broken corn), but feed the corn later in the day rather than the morning.

Keep Them Growing

You must, if the chicks are to grow, keep them 'moving' throughout their first weeks and to do this means feeding them properly, keeping them free from colds and disease and providing the young stock with sufficient exercise.

Until they are a month old they should be fed five or six times a day; until they are three months old, feed four times a day; and from then until they are adult birds, three times.

Chicks look so frail but are, in fact, surprisingly strong. Provided they are not in cold or damp conditions they will grow quickly.

Handling Poultry

Poultry should never be handled carelessly or roughly. The easiest way to catch and pick up hens is shown in Figs. 47 and 48.

Carrying live birds by their legs or wings can easily result in damage, and no more than two adult birds should be carried at one time (despite the rough handling of several birds at a time as one frequently sees in the markets). Careful and correct carrying is shown in Fig. 49.

When a bird has been caught the best method for inspecting your hen is to pass your hand under the bird's breast, from the front, and grip each thigh between two fingers. Next, raise the bird and tuck it away beneath the arm, lightly but firmly pressing the wings between your elbow and body—

the bird's head thus pointing backwards (see Fig. 50). A broody hen will be seen to have lost most of the breast feathers (Fig. 51).

Fig. 48. The correct method of picking up a bird

Fig. 47. To catch a particular hen, try to separate the bird you want from the others, corner her and close in

Fig. 49. For carrying, the bird, held in the manner above, is tucked beneath the arm. The rear should be higher than the head

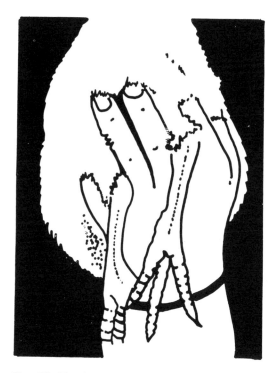

Fig. 50. The hand is slipped beneath the bird and the middle fingers between the legs as shown

Fig. 51. A broody hen shows loss of feathers from the breast

Taking a Look

The domestic poultry keeper should regularly examine his stock, if not for replacement purposes then certainly to see whether or not they are doing well. The best time to do this is at dusk when they have just settled down on the perches—first you can note the state of their crops, whether they are full or otherwise, and from this deduce whether or not you are giving the birds enough food. Before you return the birds to their perches, also examine them for lice.

If your birds, at this time of going to roost, have half empty crops, then this must mean either sickness, being out of lay, or under-feeding.

A layer in good form should be solid to the feel without undue heaviness or flabbiness—in fact, she will usually be on the spare side if anything, owing to the demands of egg laying.

Poultry for Table

Since, as already pointed out, it does not pay to keep hens after their second laying season, eating these old hens is the best bet—and you will already have had the benefit of a plentiful supply of eggs. There is another point about eating one's own poultry—even an old hen simmered for two or three hours and then browned in the oven is worth far more on one's own family table than the price you would get for such a hen.

If you decide to fatten up a few birds, then confine them as much as possible in the poultry house, otherwise the birds will run off the weight they gain. It is also helpful to cover the windows and provide semi-darkness between meals.

For fattening up purposes, whether it is an old hen or cockerels reared specially for the table, meals should preferably be of the wet mash type— plus waste milk if ever you get it—and mixed corn. Powdered milk is also a great help.

Caponising

Better results are obtained if the birds are caponised. This is the term for a cockerel into which a synthetic female hormone pellet known as oestrogen has been inserted under the skin of the neck, with the result that the male bird's testes shrink in size and activity for six or seven weeks and the bird rapidly puts on weight.

Each pellet contains 15 mg of oestrogen. Normally it is only necessary to inject one pellet, except if the birds are over six months old, then two can be given.

Some synthetic oestrogens can be included in purchased feeding stuffs, thus saving you the trouble of catching and handling the birds. With this method remember to stop feeding the hormone two weeks before killing.

How to Kill

The easiest and quickest way of killing a fowl is by dislocating its head from the neck (see Fig. 52).

In your left hand grasp the bird's legs, wings and tail and use the first two fingers of your right hand by placing them immediately behind the

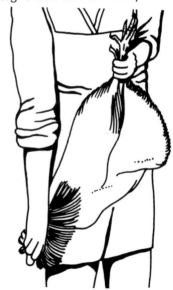

Fig. 52. Killing by dislocation of the neck

head of the fowl. The bird is then held downwards but slightly across the right leg and knee. A steady, firm pressure should be applied with your right hand until the head and neck part company. At this stage a further slight pressure is necessary to ensure that the blood vessels of the neck are also broken—otherwise you will get a very red skinned chicken of poor external quality. Although the bird will probably shake and flutter about a bit immediately after its neck has been broken, the fact is that once the spinal cord is severed the bird feels no pain.

How to Pluck

Plucking is more easily done before a bird becomes cold, since feathers come out more easily whilst the flesh is warm.

If for some reason the bird cannot be plucked at once, then it is best to leave it until quite cold—do not attempt to pluck poultry in a half cold state as the skin will be torn and the bird spoiled.

The best way to pluck a hen is to sit on a low chair holding the bird by the legs and wings—as described for killing it—with the head dangling downwards. The feathers must be plucked the **reverse** way from that in which they lie on the body.

Start on the back first, taking several feathers between your thumb and first finger of the right hand—then give a sharp pull backwards. Part of the art is to pull the feathers **quickly** to avoid tearing the skin.

After you have plucked the back of the bird, turn her over and pull the feathers from the breast and under part. Pluck the bird's neck up to within a few inches of the head, always

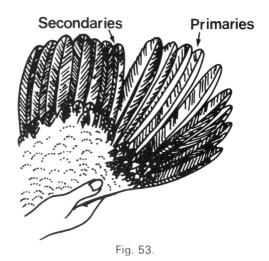

Secondaries Primaries

Fig. 53.

leaving the feathers on the top part of the neck. Next the legs should be plucked, stripping the feathers off down close to the shanks; wings next (see Fig. 53) and these take a little longer to complete owing to the larger feathers or flights which have to be pulled out two or three at a time. The tail feathers come last and these should be pulled out one at a time or the flesh may be torn.

Preparing for Oven

With the bird plucked and the head removed from the neck at the point of dislocation, lay it on the work surface on its back and take a grip between finger and thumb of the flesh just below the breastbone. With a sharp knife cut horizontally into the flesh and layer of fat to a depth of about 1 inch (25 mm), taking care not to burst the gut.

You can then put your hand into the carcass and draw out the entrails. From what you remove in this piece of bird surgery, the heart, liver, outer case of the gizzard (that is a second stomach where the bird grinds its food) and the remainder of the neck are known as the giblets—very nice for simmering down into a broth.

When you are drawing a bird try not to break or tear any of the organs but, should this happen, wipe out the mess with a damp cloth or paper towels—it is unwise to wet the carcass by washing unless the bird is being cooked immediately (Figs. 54 to 57).

The job is complete when you have pulled off the remainder of the neck, folded the skin over the resulting hole, and trussed the bird by tying the 'drumsticks' to the 'parson's nose'. Do this in a tidy manner and you will be justifiably proud of your art.

The home quick freezer is particularly useful for storing oven-ready poultry, either whole or in pieces. The birds must, of course, first be 'dressed' (i.e.,

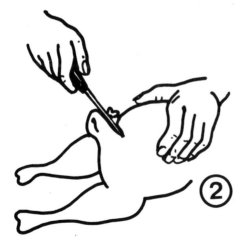

Fig. 54. The neck is cut off as close to the shoulders as possible

Fig. 56. The vent is loosened by cutting around it—be careful not to cut into the intestine. The entrails are removed through a short, horizontal cut about 5 cm below the cut made around the vent

Fig. 55. The oil sac on the back near the tail should be cut out, as it sometimes gives a peculiar flavour to the meat. Remove it with a wedge-shaped cut

Fig. 57. A well-trussed bird is a pleasing sight. In the method shown here the legs are placed under the strip of skin between the vent opening and the cut from which the intestines were removed

internal organs removed as described above) and placed in moisture-resisting wrappers before freezing.

Chicken should not be stuffed before freezing as this limits the storage life of the bird. (A stuffed chicken should be eaten within a month.) Store the giblets, wrapped separately, beside the bird.

Joints of chicken take up less freezer space than whole birds. To joint a fowl for the freezer, cut it in half through the breast bone and back bone—then cut each half into two or three pieces. Pack on an aluminium or fibre food tray and, finally, wrap with foil or polythene (see Fig. 58).

The quick freeze process works on the principle that the product is frozen so rapidly that no internal changes—for example, loss of moisture—can take place. The freezing is done rapidly at temperatures ranging from about −18°C to −34°C with the result that the natural quality, flavour and appearance of the birds is preserved.

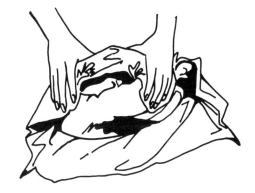

Fig. 58. Wrapping a chicken

This can be done with foil—as in the top sketch or as in the lower picture the bird can be placed in a polythene bag and secured with a clip

Preserving Eggs

It is possible that during the 'flush' season in the spring and early summer you will get more eggs each week than the family needs. Surplus eggs can be preserved and kept for use in the winter, provided you observe a few necessary but simple principles.

Make sure that eggs for storage are absolutely new laid and, if possible, put them in the preservative the very day you collect them from the nest boxes (second day should be the limit). The most generally used preservative is called water-glass, the name commercially applied to a solution of silicate of soda. A small tin of water-glass (half a kilogram), obtainable from chemists' shops, is sufficient to preserve 50 eggs.

Dissolve this quantity of water-glass in one gallon of hot water that has been well boiled (this is to kill any bacteria which may be present). When the solution is cool pour onto the eggs, which should be placed in the container (an enamel bucket is ideal) with their smaller ends pointing downwards. Make sure that the topmost layer of eggs is well covered—this also applies, of course, if you continue to put a few eggs in from day to day during the 'flush' season. Keep the container free from dust by covering with a lid and do not store your preserved eggs in a temperature which falls to freezing point or which rises above about 5°C.

Choosing the Best

The deteriorative changes to which the contents of an egg are liable are primarily caused by exposure to air and warmth. As the absence of sufficient warmth to cause deterioration can only be effectually secured by commercial cold storage methods, the home poultry producer achieves a similar result with water-glass—it simply closes the pores of the eggshell and so excludes air.

When preserved eggs are taken from the water-glass solution for use they are coated with a sticky white covering and must therefore be washed in warm water before use. Another hint is to prick the shell with a needle to prevent cracking when you boil the eggs.

From March to June is generally the best period in which to 'put eggs down'. Eggs laid in the hot days of summer are much less suitable. Eggs for preserving must be clean but unwashed, with firm and smooth shells; preferably infertile. They will keep at least 10 months under cool conditions.

Uses of Poultry Manure

The output of fresh manure per hen per day is over 4 oz. (90 g)—in fact, if you had 100 birds you would make nearly four tons a year!

Poultry manure is far richer in sulphate of ammonia, super phosphate and potash salts than farmyard manure. Its richness is due to the large amount of plant food elements that pass quickly through the fowl, and to its content of semi-solid urine, the white part of the droppings, with its nitrogen content.

As it is rapidly available, and much of the value can be washed away by rain, poultry manure is best applied in the spring. Use it in the vegetable plot and for your flowers but it is not recommended for apples and pears, although it is certainly most useful for soft fruit growing.

Good for Tomatoes

The standard dressing of poultry manure for the vegetable garden and the flower borders is about 4 oz (90 g) to the square yard (metre) forked in during February and March—prefer-ably when you have limed the garden the previous Autumn. Never mix lime and poultry manure, or dig them in within three months of each other, or both will be wasted.

During the Summer you can use your poultry manure as a quick acting tonic. It can also be used with tomatoes, potatoes and onions, but in these cases it requires a 'balancer' dressing of potash.

Better Compost

Poultry manure also helps you to make more and better compost.

If swift heating of compost is to be secured there must be a great increase of bacteria in a short time and these require readily available nitrogen and phosphorus and this is where your poultry manure comes in.

One good way to acquire a supply of this valuable material is to build a heap consisting of a base of compost, followed by an inch (20 to 30 mm) of poultry manure, then more vegetable waste, followed by a layer of 2 inches (50 mm) of soil (plus a sprinkling of lime), more compost material and then back to your poultry manure.

Suitable Breeds

Part of the pleasure of domestic poultry keeping is to know something about the various breeds of poultry and how to distinguish them—and, of course, making your own particular choice. Breeds are very distinctive, not only in colour and—to some extent—shape, but also in features such as the feathers and combs (see Figs. 59 and 60).

First, some of the deservedly popular

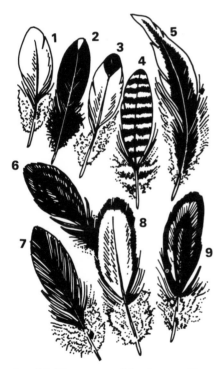

Fig. 59. Nine types of feather marking

(1) Self, (2) Tipping, (3) Spangling, (4) Barring, (5) Striping, (6) Pencilling, (7) Peppering or Smut, (8) Single, and (9) Double Lacing

'light' breeds, all of which will lay you plenty of eggs even if they do not make quite such a weighty table bird when they reach the boiler stage:

Fig. 60. Types of comb

For Town Conditions

Leghorns: Black, brown and white— the latter being the most popular (Fig. 61). These birds are useful all-purpose poultry, and they do well in towns. Robust and hardy. White eggs.

Leghorns were long known in Italy, but it was not until about 1835 that the breed first spread to America and later to us. No matter what particular colour group, this breed of hen not only lays well but does not readily go broody. The cockerels make passable table birds—but only just.

Perhaps the main advantage of Leghorns is that they are particularly adaptable and do not seem to mind

Fig. 61. White Leghorn hen and cock

Fig. 62. Pair of Black Minorcas, bantam weight

being kept in a confined space. Without any doubt, the Leghorn is the outstanding champion egg layer of all breeds.

Ancona: None of the other light breeds is as well known as the Leghorn race although the Ancona has its enthusiasts. Again a breed imported from Italy and of a very pleasing speckled plumage. White eggs.

The true Ancona is a prolific layer and a very thrifty feeder, requiring no extra food to persuade them to continue laying in the winter.

No other hen, except perhaps the Derbyshire Redcap—now a rare variety—is as active as the Ancona, and it is not a bird to go broody easily. If you live in the country, for six months or more it will require very little food since these birds forage for themselves in the hedgerows and fields.

Minorca: Another non-sitting fowl and one of the oldest breeds of all. Its origin is obscure but there can be little doubt that the first of these fowl found their way to Britain from Mediterranean islands.

The Minorca (Fig. 62) is a sprightly and alert little bird but less popular than it was, except among some of the fanciers. These black plumaged birds lay a good-sized white egg.

Of the light breeds listed above only the Leghorns are used to any extent in commercial production of crossbred fowl, mainly being mated with Light Sussex or Rhode Island Reds of the heavier breeds. These first crosses make excellent domestic poultry.

The Bigger Birds

Rhode Island Red: The most popular of the 'heavy' breeds, this fowl (Fig. 63)—as the name implies—came to us from Rhode Island, USA. The breeder's object was to produce the most eggs and the best table fowls at the least expense and trouble. Brown eggs.

The domestic poultry keeper is likely to find the Rhode the quietest and most easily managed of all poultry —an old Rhode Island hen is often more like a domestic pet than part of the poultry flock.

The Rhode Island Red, a handsome creature in its deep brownish red plumage, is a very much general purpose breed—the average male weighs around 8 lbs and in laying tests the Rhode usually exceeds 200 eggs a year.

Wyandottes: Another of the better known general purpose breeds (Fig. 64). They are among the most popular fowls in the world and some varieties of Wyandottes are highly colourful poultry with varieties known as Gold-Laced, Silver-Laced, White, Buff, Gold-Pencilled and Silver-Pencilled.

Again a breed of American origin, and Wyandottes are certainly of striking beauty—hence being popular with domestic poultry keepers who also want to exhibit.

Of the various types of Wyandotte, the White Wyandotte is probably most often seen in this country. All Wyandottes are very hardy and good winter layers but in some strains their white eggs are not quite so large as those of the Rhode Island Red or the Light Sussex.

Sussex Fowl: This is a very old English breed, the most popular of which is the Light Sussex (Fig. 65), a white feathered bird with black tail feathers and black hackles (neck feathers).

Fig. 63. Rhode Island Red pullet and cock, heavy

Fig. 64. White Wyandotte hen and cock

Fig. 65. Light Sussex cock and hen

Sussex poultry are probably descended from what were called, a century ago, Southern Old English Fowl. They were found throughout Kent, Sussex and Surrey and the first Sussex poultry club was founded at Lewes over 70 years ago.

Hardy, good looking birds laying tinted coloured eggs, they grow rapidly and make an excellent table fowl—cockerels often exceeding 9 lbs (4 kilos) in weight.

Hybrids

Apart from these leading, pure-bred 'heavies' there are a number of first-rate modern hybrids (as well as hens produced from first-crosses) which the amateur might wish to consider buying as young stock. All hybrids are very hardy and good layers.

The new hybrids have been produced by selective breeding, using strains which records show to be above average layers. In-breeding then stabilises a strain so that all those in one 'family' have similar characteristics. After long and painstaking tests the breeders decide on the best hens for crossing with the selected 'family' and production begins on a commercial scale.

Whilst the professional breeders have succeeded in improving laying and meat producing qualities, hybrids are primarily produced for the commercial poultryman who operates under ideal, controlled conditions of feeding, housing, lighting and the rest. **Good as the traditional breeds are, they are not now so easily obtained as hybrids which in any case will give you excellent results.**

The Fancier

This is the term used for poultry keepers who breed primarily for exhibition of 'fancy breed' points.

Fanciers may be divided into three sections: (1) those who breed and exhibit for the pleasure it affords—even if it costs money; (2) those who breed and exhibit entirely for business purposes; and (3) the amateur and beginner—the genuine hobbyists of limited means who regard keeping fancy poultry as an interesting pursuit —like pigeons—with the possibility of making it pay a small profit, usually by selling pedigree birds.

The fancier aims for the standard characteristics of the bird of his choice and always eliminates poor specimens. Though he may also keep the birds to lay eggs there is no question of choice for him between a good breeder and a poor layer, or a poor breeder and a good layer, since the fancy points are the more valuable to this type of amateur poultry keeper.

Small is Beautiful

Quite frequently domestic poultry keepers specialise in the Lilliputian breeds of poultry called Bantams. These little hens take up only a small amount of space and are certainly economical to feed. The egg they lay is smaller, but Bantam eggs are great favourites with the children.

It is generally thought that the original Bantams came from Java. These perky little birds have a different action in walking from that of the standard hen—a kind of jerky gait which gives them a jaunty manner. It is, however, a mistake to think that Bantams can be kept more easily than larger breeds of poultry. In fact, the contrary is true since in some respects Bantams require more attention than large fowls, especially if they are kept for exhibition, and they are more prone to colds.

Bantam fowl eat only about half the quantity of food needed by a standard hen, but neither do they give so many eggs. They average about 100 eggs a year, although individual strains of Bantam have laid as many as 175 eggs per bird each year.

There are Bantams of most varieties of full-sized poultry, including Rhode Island Red Bantams and a scaled-down version of that famous old breed the Plymouth Rock (see Fig. 66).

A suitable house for six Bantams need be no larger than 4 feet 6 inches (1.5 m) long by about 3 feet (1 m) wide, with a height of just under 3 feet

Fig. 66. Barred Plymouth Rock bantam

(1 m) in front, sloping 4 inches (90 mm) to the back. To this house attach a run about 9 feet (3 m) long and the same width.

Popular Game Birds

If there is one breed of fowl that belongs more to England than any other, it is the Game fowl. At the time of the Roman occupation of our land it is recorded that the Britons kept fowls 'for pleasure and diversion' and although the breed is not specifically mentioned, or the sport defined, no doubt the historian was referring to cock fighting. Some of the earliest Chinese records also mention cock fighting, while in India there are references to the same thing going back to 1000 BC.

Today, Game birds are among the most popular of all Bantam type poultry. They have a large number of admirers and, like all Bantams, make interesting birds for the family poultry keeper whether he exhibits at shows or not.

Apart from their historical association with the now banned cock fighting, Game fowl have provided the small poultry fancier with a fascinating hobby and the best materials with which to try his skill at breeding for exhibition purposes.

Game fowls are divided into two classes—Old English and Modern. Both are composed of many varieties, each named according to the colour of its plumage. Modern Game fowl mainly owe their origin to the warriors of the old-time cockpits and are characterised by their long legs and heads and shortness of feather. It was, however, devoid of economic utility qualities and the fanciers turned to the more useful Old English type of fowl (see Fig. 67)—the perfect combination of the useful and beautiful in bird life.

Game birds lay white shelled eggs which, although a little below average in size, are unequalled for fullness of yolk and richness of flavour.

Fig. 67. Old English Game Spangled cock

Ducks

So economically may ducks be kept, and so good is the return they give, that many domestic (and commercial) poultrymen prefer them to fowls.

Ducks are the cheapest form of livestock to house and if you keep a laying breed such as the Khaki Campbell (as distinct from a table breed like the well-known white Aylesbury) they lay more eggs than fowls—particularly so in the second year of production (as many as 300 eggs a year is not uncommon). Kept as **table** birds, ducks attain killing size much quicker than chickens—at 10 weeks an Aylesbury can weight at least 5 lbs (2 kilos) and they can be fed very largely on potatoes and greenstuff.

There is just one snag for the domestic keeper: Ducks are not recommended for small back gardens in built-up areas—their quacking and the quagmire they make of a small run in wet weather are serious disadvantages. But if your house stands in the open and has a large garden, or access to a field or grass run, as many village properties do, then ducks are both profitable and a joy to own. Swimming water is not necessary unless you keep ducks for breeding, when a pond is desirable if the ducks are to mate successfully and ensure good fertility.

Housing for Ducks

You may hear it said that ducks can live an entirely open air life. Maybe, but much the best results are obtained with both laying and table ducks by housing the birds at night and also making sure the ducks have daytime shelter from hot, summer sun by providing enough shade from trees and shrubs. But certainly the housing for ducks is far less elaborate than that needed for hens. There is no need for a house with glass windows and other fittings; all ducks need is a dry, well-ventilated house giving them around 4 to 5 sq feet (.5 sq m) of floor space each—so a simple house 4 feet 6 inches (1.5 m) long by 4 feet (1.25 m) wide will provide very ample room for a flock of 10 ducks. Such a house need only be about 4 feet (1.25 m) high at the front and just under 1 yard (1 m) at the back—ducks do not perch, as hens do, and this is why the structure does not need to be as high as a hen house.

A slatted wooden floor, which helps to keep the litter dry, should be fitted to the ducks' house and raised about 2 inches (5 cm) off the base. The upper 1 foot (.3 m) of the front should be filled in with wire netting, protected by an overlaying weatherboard.

Make the whole front loose, or on hinges, so that it can be removed bodily, or opened, in warm weather.

Ducks are clumsy creatures, so do give them a sloping run-up board into the house otherwise if they have to jump to get inside they will injure themselves. Nest boxes are simple affairs of straw on the floor, kept in position either with a few bricks or some shallow planks of wood.

If possible, face a duck house towards the south-east, south or south-west to get all the sun possible and have a small run attached—this is because ducks should be kept penned up till about 10 a.m., by which time eggs will have been laid in the house instead of wasting your time searching the grounds later in the day.

Feeding

As much as three-quarters of a duck's daily food ration can consist of vegetable matter. Cooked potatoes, carrots and swedes are the mainstay and these should be mixed with water and dry mash (obtainable, like layers' mash for hens, from corn merchants) to form a fairly sloppy but appetising meal. Two meals a day should be given, allowing each duck a little over 4 oz (100 g).

As with hens, supplies of water, grit and oyster shell must always be available for ducks. Although, as already said, swimming water is not essential for a domestic flock of ducks kept only for meat or egg production, the birds being waterfowl means they must have water deep enough to immerse their entire head to keep healthy.

If your ducks are to average 300 eggs a year—and this should be your target per bird—they must be well looked after even though the labour required is ridiculously small for such a hefty return.

Routine work consists of loosening up the floor litter **every** morning and changing the litter as soon as it becomes wet through and foul. While straw will last only a couple of weeks before it needs changing, peat moss litter will last two **months** if kept raked over.

Ducks do not like fierce winds—egg yield can be cut by a third if the birds lack shelter. A hedge, a row of trees, a clump of shrubs, a range of buildings are all satisfactory places for shelter—and if you cannot manage any of these then a few sheets of corrugated iron placed lengthways will serve the purpose.

Above all, remember that these delightful birds respond to tranquillity in every way. They need quiet and gentle handling, and a move to a different house—or a change of feeding—when ducks have begun winter laying will not only reduce egg yields but probably put some into partial moult.

Rearing Ducklings

Ducklings is the term applied to young ducks and drakes. They can, like chickens, be reared either by natural or artificial means.

By broody hen: Make a comfortable nest on a sod of turf—remembering to moisten around the nest in very warm or dry weather, since duck eggs require more moisture than hen's eggs. During the 28 days needed for incubation manage the 'mother hen' just as described for hatching chickens under a broody.

By brooder: Ducklings need a little less heat than chickens, and any ordinary good brooder (as described on page 46) will rear ducklings, but the litter beneath the brooder will need changing frequently as it speedily becomes soaked through. Up to the third day ducklings need to be watched closely to make sure they know, after wandering about, where to go back for warmth (they can be annoyingly stupid little creatures!).

Picking out the best layers is as important with ducks as with hens. Feeding stuffs are too dear to waste on below-average birds—so learn to know the characteristics of the good laying duck; she is active, first at the food container and last away, and is early 'about the house' in the mornings and late to 'get into bed' at night!

Another sign of a prolific layer is the appearance of being nice and plump behind and walking with that typical duck manner of legs wide apart. The good duck is all.curves; an angular looking bird is a doubtful layer.

Breeds to Keep

It is generally agreed that all domestic ducks (with the exception of the Muscovy from South America) are descended from the wild duck or Mallard.

The following are the most important of the very large number of different breeds of domestic duck:

Aylesbury: This duck, originating from the Vale of Aylesbury, rightly occupies the top position among birds bred for the table. Its popularity stems from its ability to attain size and weight at a very early age—good alike for the early duckling trade and the home food producer. Quality and flavour of flesh is excellent. Lays a large bluish-green egg. Plumage, glossy pure white (Fig. 69).

Khaki-Campbell: Developed from the original bird named after a certain Mrs Campbell who crossed an Indian Runner (her particular bird laid 182

eggs in 196 days!) with one of a very hardy Rouen breed. The modern Khaki-Campbell is **the** outstanding egg layer, frequently giving 300 white eggs in one year. Plumage of the drakes is khaki colour all over, except for bronze-green head and stern; the ducks are entirely khaki colour with a few feathers of lighter shade on back and wings (Fig. 70).

Indian Runner: Formerly classed as the best egg laying breed of duck. There are five varieties—black, chocolate, fawn, fawn and white, and white.

Of Indian origin, this breed was introduced into the county of Cumberland about 100 years ago. The bird is of a different formation from other ducks—with a high and erect shape to its body and a flat skull. Eggs are medium-sized and white.

Cayuga: Not as popular as it should be, considering the breed combines good laying and table qualities. These birds are a deep greenish-black in colour and have also been known as the Black Duck of North America (it is noteworthy that a lake in North America bears the name Cayuga and black coloured duck abound there). The shape of the Cayuga duck is like that of the Aylesbury, but its eggs are dark green.

Rouen: A complete contrast in egg colour from the breed mentioned above —the Rouen lays a **blue** egg. It is an excellent table bird, often crossed with the Aylesbury. The Rouen's plumage is very like that of the lovely wild Mallard (the reason why so many of the breed are seen on ornamental waters) and the Rouen male undergoes the same peculiar changes in summer when the drake's beautiful bottle-green head and brownish-red markings are lost in the moult and he becomes nearly indistinguishable from the drab-ber, brown-mottled duck.

Pekin: Imported from China in the 19th century, this excellent table bird is second only to the Aylesbury, and as a layer of eggs it is **better** than the latter—hence Aylesbury and Pekins are often crossed. The Pekin breed of duck is very hardy and a thrifty forager for food. The colour is a uniform cream throughout, with bright orange bill, shanks and feet. Pekins lay a blue egg (Fig. 68).

Orpington: For some years, when the rage for buff colour in all poultry was at its height, these ducks enjoyed a considerable vogue but they are less popular today. However, they are a useful dual purpose breed ideal for limited space. Orpingtons make very presentable ducklings, to eat or to sell, at eight to 10 weeks and laying qualities are above average for a recognised table breed.

Colour of this breed is either buff or blue and white. They are excellent free range birds, and they lay a white egg.

Fig. 68. Pekin duck

Fig. 69. Aylesbury duck

Fig. 70. Khaki-Campbell duck

Know your Law

In addition to the necessary check with local authority regulations regarding poultry keeping referred to early in this book, some basic knowledge of the law as it touches on livestock keeping is worth having.

Trespass. Poultry are proverbial trespassers and often the source of friction between neighbours. In this respect the law generally applied to the ownership of animals also covers poultry—and this says that owners must keep their animals on their own land and if they do not do so they will be liable to compensate neighbours for any damage done. So, if your poultry are not properly fenced in and they enter your neighbour's garden and ruin his immaculate seed bed, scratch out his prize potted plants or peck his fruit then **you** will be liable because you neglected to perform your duty of providing adequate fencing-in.

A mistake frequently made is to believe that the owner of a fence dividing two properties is the person whose duty it is to keep and maintain the fence in such condition as to make trespass impossible. It is true that in certain cases owners of fences in the country are under what is called a 'prescriptive obligation' to maintain such fences for their neighbour's benefit but there is never any such obligation where **poultry** is concerned.

Supposing the worst has happened and your neighbour is irate owing to your poultry being in his garden! In order to obtain **damages** your neighbour must furnish **proof** both of the accuracy of his costing of damage done and **proof** that this damage was, in fact, caused by your poultry not being properly fenced in.

It is in law open to the poultry keeper to show that he had taken 'all reasonable precautions' and that it was by reason of some **third party's act** that the fowls escaped (someone leaving a gate open, for instance). If that was so, then the poultry keeper would not be liable. In order for the neighbour to obtain an injunction it would be necessary for him to show that you, as the poultry keeper, threaten or intend a repetition of the trespass or that there is reason to fear that it will happen again.

Unlike ducks and geese, hens have no right of way on the road. If a motorist kills them he is not obliged to pay.

Dogs and Cats. Of particular interest to poultry keepers is the question of damage to poultry by dogs and cats.

Here the poultry owner is covered in the same way as a farmer and his livestock whereby liabilities are imposed on the owner of a dog for injuries done to cattle. In proved cases of other people's pets injuring or even killing your poultry compensation is recoverable.

Law of Nuisances. There are several ways in which poultry might land you in trouble as regards what the law would define as a nuisance. Trespass (see above) is a nuisance—so, too, is cock crowing or the keeping of poultry under insanitary conditions.

Cock crowing is an offence not against any particular Act of Parliament, but against local bylaws which, in fact, have the same effect. Some local authorities have bylaws providing that 'cattle, dogs and poultry shall not be kept in such places or in such manner as to be a nuisance or annoyance to the inhabitants'.

Insanitary Conditions. All alleged nuisances under this heading can be dealt with under the Public Health Acts. Complaint about your poultry must first be lodged to the local authority who will then, if satisfied with the genuineness of the complaint, serve notice on the poultry keeper to stop the nuisance within a certain time—or be taken to court.

Cruelty. As with other livestock reared for human consumption, poultry have to be killed for food and no offence of cruelty is committed unless the killing is accompanied by inflicting unnecessary suffering. Correct methods of killing have already been described in this book (see page 50) and the newcomer to poultry keeping can study the leaflets on this and other aspects of domestic poultry keeping issued by the Royal Society for the Prevention of Cruelty to Animals (send a stamped and addressed envelope to the Society at the Manor House, Horsham, Sussex).

Most important of all, obtain from H.M.S.O., a copy of a Ministry of Agriculture leaflet entitled *Code No. 3 Domestic Fowls*. This covers official policy on the welfare of poultry.

Glossary of Poultry Terms

Air Cell   Air space usually found in the large end of the egg.

Albumen   The white of the egg.

American Breeds   Those breeds developed in America and having common characteristics such as yellow skin, non-feathered shanks, red ear lobes.

Axial Feather   The short wing feather between the primaries and secondaries.

Baby Chick   Newly hatched chick before it has been fed or watered.

Bantam   Diminutive fowl. Some are distinct breeds, others are miniatures of large breeds.

Barring   Alternate stripes of light and dark across a feather, most distinctly seen in the Barred Plymouth Rock breed.

Beak   Upper and lower mandibles of chickens, turkeys, pheasants, pea fowl, etc.

Bean   Hard protuberance on the upper mandible of water fowl.

Beard   A bunch of feathers under the throat of some fowls, such as Faverolles, Houdans and certain varieties of Polands.

Bill   The upper and lower mandibles of waterfowl.

Blood Spot   Blood in an egg caused by a rupture of small blood vessels usually at the time of ovulation.

Boots   Feathers projecting from the toes, as in the Brahma and Cochin breeds.

Breast   The forward part of the body between the neck and the keel bone.

Breed   A group of fowl related by ancestry and breeding true to certain characteristics such as body shape and size.

Broiler   Young chickens under 9 weeks of age of either sex, that are tender-meated with soft, pliable, smooth-texture skin.

Brooder   Heat source for starting young birds.

Broody   Maternal instinct causing the female to want to hatch eggs.

Candle   To determine the interior quality of an egg through the use of a special light in a dark room.

Cannibalism   The habit of eating other birds in the flock.

Cap   A comb; also the back part of a fowl's skull.

Cape   The feathers under and at the base of the neck-hackle, between the shoulders.

Capon   A male fowl treated with a female hormone to produce better quality table birds.

Cock   A male bird over 12 months of age.

Cockerel   A male bird under 12 months of age.

Comb   The fleshy prominence on the top of the head of fowl.

Crest   A crown or tuft of feathers on the head; sometimes called the 'top-knot' and known in Old English Game as the 'tassel'.

Crop   An enlargement of the gullet where food is stored and prepared for digestion.

Crossbred	The first generation resulting from crossing two different breeds or varieties.
Cull	A bird not suitable to be in a laying or breeding pen.
Culling	Removing unsuitable birds from the flock.
Debeak	To remove a part of the beak to prevent feather pulling or cannibalism.
Drake	Male duck.
Dub	To trim the comb and wattles close to the head.
Ear Lobe	Fleshy patch of skin below ear. It may be red, white, blue or purple, depending upon the breed.
Embryo	The young organism in the early stages of development, as before hatching from the egg.
Face	Skin around and below the eyes.
Flight Feathers	Primary feathers of the wing, sometimes used to denote the primaries and secondaries.
Fowl	Term applied collectively to chickens, ducks, geese, etc. or the market class designation for old laying birds.
Gander	Male goose.
Germinal Disc or Blastodisc	Site of fertilisation on the egg yolk.
Gizzard	Muscular stomach. Its main function is grinding food and partial digestion of proteins.
Goose	The female goose as distinguished from the gander.
Gosling	A young goose of either sex.
Gullet or Oesophagus	The tubular structure leading from the mouth to the glandular stomach.
Hackle	Plumage on the side and rear of the neck of fowl.
Hen	A female fowl more than 12 months of age.
Hock	The joint of the leg between the lower thigh and the shank.
Horn	Term used to describe various colour shadings in the beak of some breeds of fowl such as the Rhode Island Red.
Hover	Canopy used on brooder stoves to hold heat near the floor when brooding young stock.
Isthmus	Part of the oviduct where the shell membranes are added during egg formation.
Keel Bone	Breast bone or sternum.
Laced, Lacing	A stripe or edging all round a feather, differing in colour from the rest of the feather.
Leg Feathers	Feathers projecting from the outer sides of the legs, of such breeds as Brahmas, Cochins and other fanciers' breeds.
Litter	Soft, absorbent material used to cover floors of poultry houses.
Magnum	Part of the oviduct which secretes the thick albumen or white during the process of egg formation.
Mandible	The upper or lower bony portion of the beak.
Marking	The barring, lacing, pencilling, spangling, etc. of the plumage.
Moult	To shed old feathers and regrow new ones.
Mossy	Confused or indistinct marking: A defect.
Muffing	The beard (as in Faverolles, for instance) and the whiskers, i.e. the whole of the head feathering except the crest.
Oil Sac (Uropygial Gland)	Large oil gland on the back at the base of the tail—used to preen or condition the feathers.
Ova	Round bodies (yolks) attached to the ovary. These drop into oviduct and become yolk of the egg.

Oviduct	Long glandular tube where egg formation takes place and leading from the ovary to the cloaca. It is made up of the funnel, magnum, isthmus, uterus and vagina.
Pencilling	Small markings or stripes over a feather—straight across in Hamburgh hens, and often known as bands.
Pendulous Crop	Crop that is usually impacted and enlarged and hangs down in an abnormal manner.
Perch	A wooden pole on which fowl rest or sleep.
Plumage	The feathers making up the outer covering of fowls.
Poult	A young turkey.
Poultry	A term designating those species of birds which are used by man for food and fibre and can be reproduced under his care. The term includes chickens, turkeys, ducks, waterfowl, pheasants, pigeons, peafowl, guineas and ostriches.
Primaries	The long stiff flight feathers at the outer tip of the wing.
Pubic Bones	The thin terminal portion of the hip bones that form part of the pelvis. Are used as an aid in judging productivity of laying birds.
Pullet	Female chicken less than 1 year of age.
Recycle, or Force Moult	To force into a moult with a cessation of egg production.
Replacements	Young birds which will replace an old flock.
Roasters	Chickens of either sex, usually 3 to 5 months of age.
Roosting	Fowl at rest or sleeping.
Rose Comb	A broad, solid comb, nearly flat on top, covered with several small regular points and finishing with a spike. Seen very well in the Redcap breed, long familiar in Derbyshire and South Yorkshire.
Saddle	Lower part of the back.
Secondaries	The large wing feathers adjacent to the body, visible when the wing is folded or extended.
Sex-Linked	Any inherited factor linked to the sex chromosomes of either parent. Plumage colour differences between the male and female progeny of some crosses is an example of sex-linkage. Useful in day old sexing of chicks.
Shank	Leg.
Shell Membranes	The two membranes attached to the inner egg shell. They normally separate at the large end of the egg to form air cells.
Sickles	The long, curved feathers of a cock's tail.
Snood	Fleshy appendage on the head of a turkey.
Spangling	The marking produced by a large spot of colour on each feather differing from that of the ground colour.
Spur	The stiff, horny projection on the legs of some birds. Found on the inner side of the shanks.
Strain	Fowl of any breed usually with a given breeder's name and which has been reproduced by closed-flock breeding for several generations.
Tail Coverts	Soft, curved feathers at the sides of the lower part of the tail.
Tail Feathers	Straight and stiff feathers of the tail only. (In the male fowl the tail feathers are contained inside the sickles and coverts.)
Testes	The male sex glands.
Thigh	That part of the leg above the shank.

Tom	A male turkey.
'Top-Knot'	(See Crest.)
Trachea or Windpipe	That part of the respiratory system that conveys air from the larynx to the bronchi and to the lungs.
Under colour	Colour of the downy part of the plumage.
Uterus	The portion of the oviduct where the thin white, the shell and shell pigment are added during egg formation.
Vagina	Section of the oviduct which holds the formed egg until it is laid.
Variety	A sub-division of breed—distinguished either by colour, colour and pattern or comb type.
Vent	The external opening, the anus.
Wattles	The thin pendant appendages at either side of the base of the beak and upper throat, usually much larger in males than in females.
Wry Tail	A tail carried awry, to either side of the continuation of the backbone.
Yolk	Ovum, the yellow portion of the egg.

Other Books in Print

Poultry Nutrition. HMSO 1963
Poultry Culture for Profit (Sturges). Spur Publications 1975
Poultry Production (Card and Neisheim). Bailliere 1973
Natural Poultry Keeping (Worthington). Crosby Lockwood 1970
Modern Poultry Keeping (Portsmouth). Teach Yourself Books 1965
Diseases of Poultry (Hofstad). Bailliere 1972